NULL OBJECT
Gustav Metzger thinks about nothing

Bruce Gilchrist, Jo Joelson
London Fieldworks

black dog
publishing
london uk

CONTENTS

A deep distrust of blindly following my thoughts any farther in this
direction suddenly crept over me. I stretched out straight in bed and
covered my eyes and ears with my hands so as not to be distracted
by my senses; so as to kill off every thought.

But my determination was smashed by an iron law: one thought
could only be driven away by another thought, and if that one should
die there would already be the next feasting on its flesh. I sought
refuge in the roaring torrent of my blood, but my thoughts were
ever at my heels; I hid in the pounding forge of my heart, but after
a short while they had discovered me there.

The Golem
Gustav Meyrink, 1915

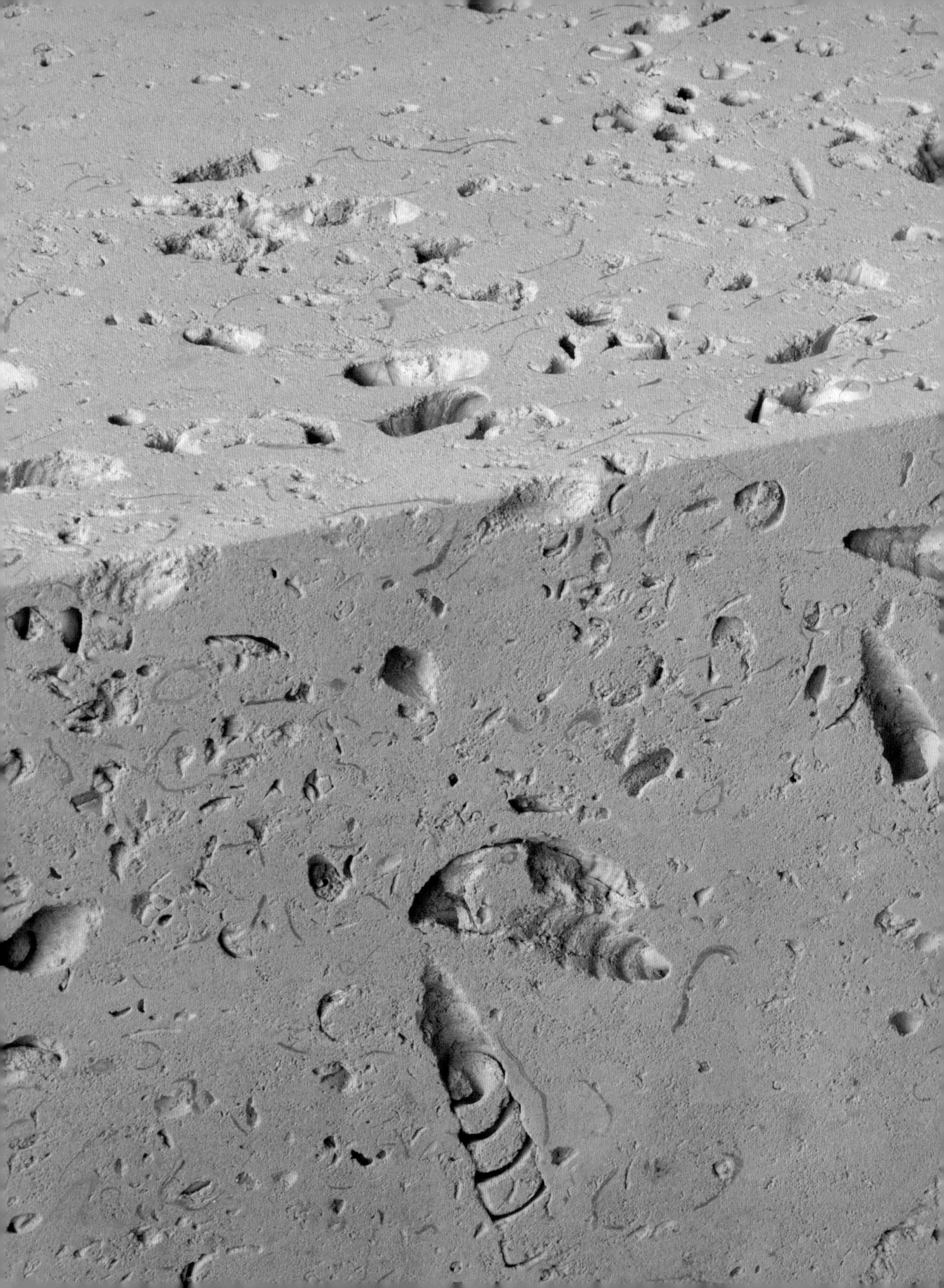

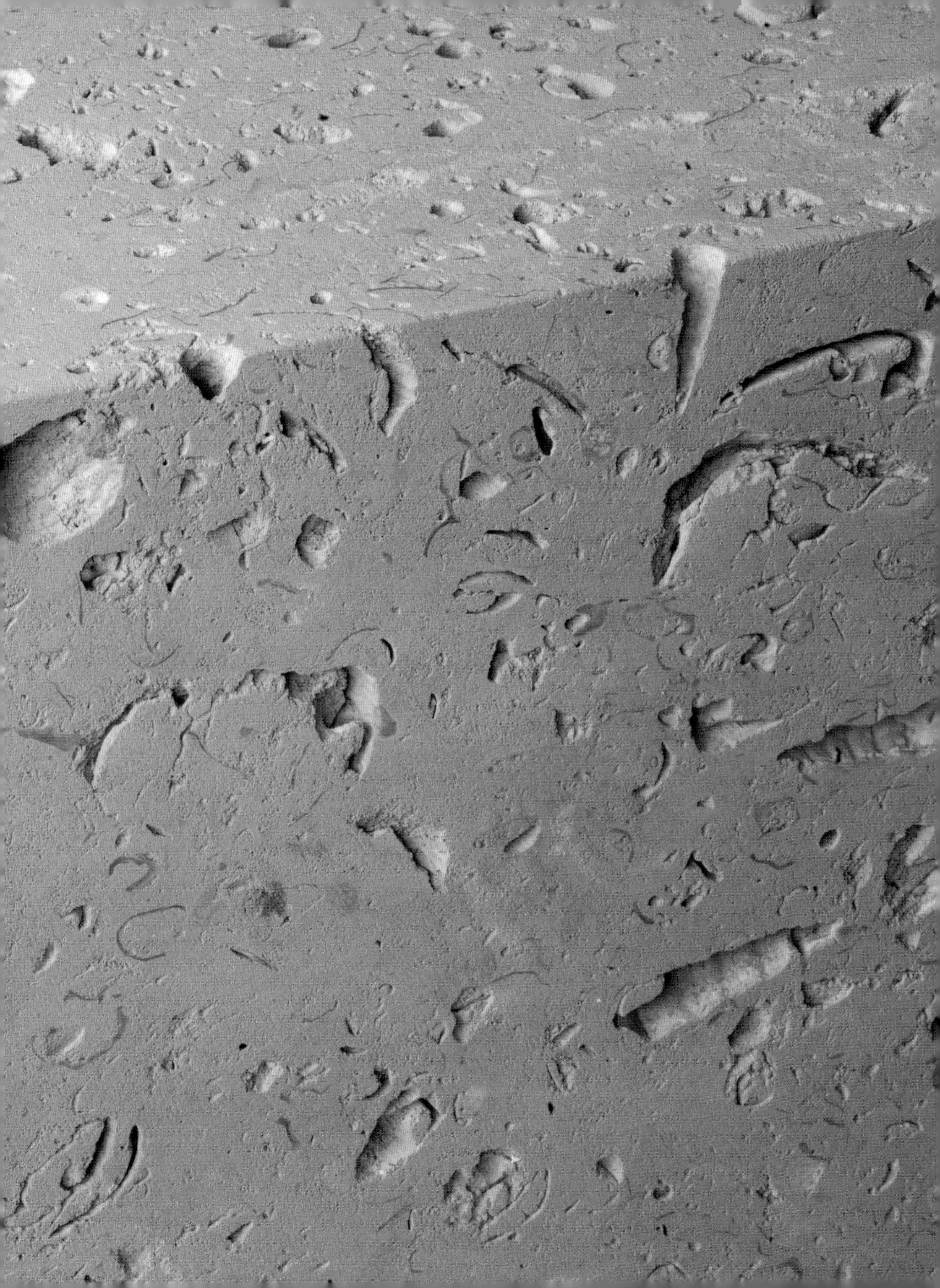

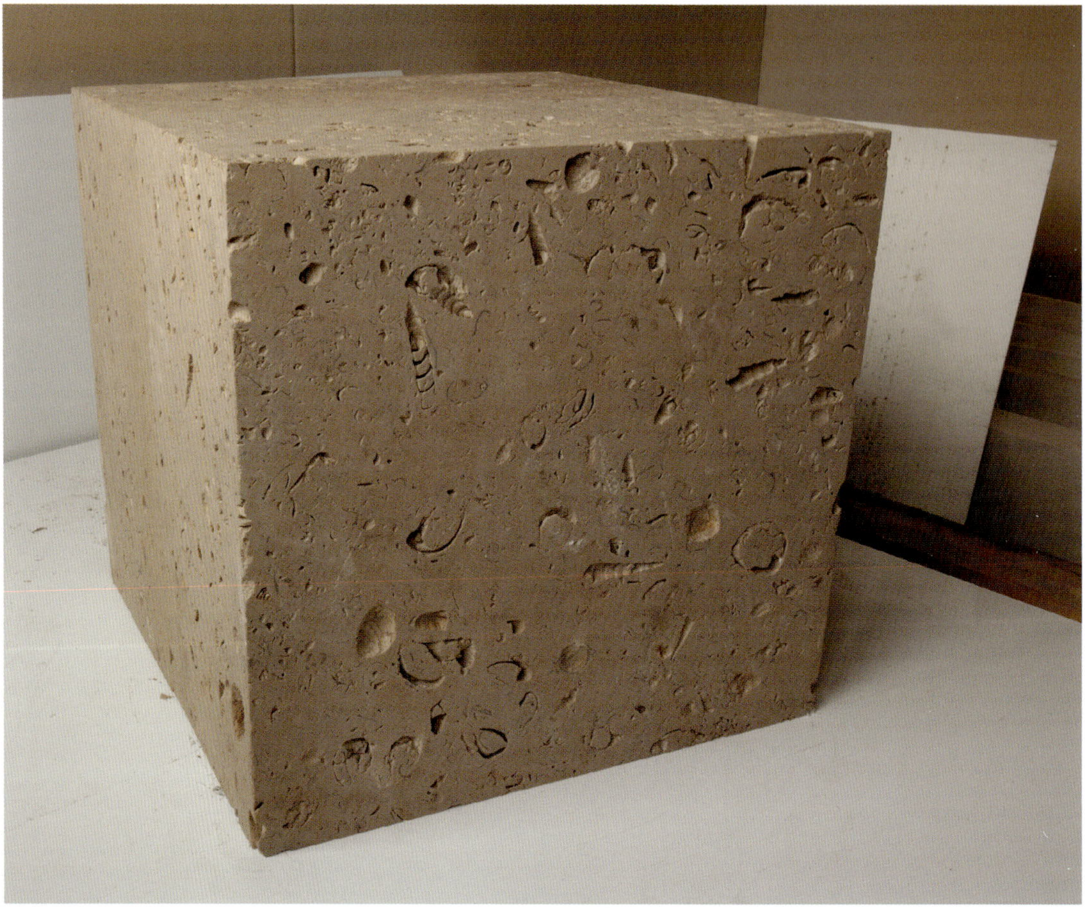

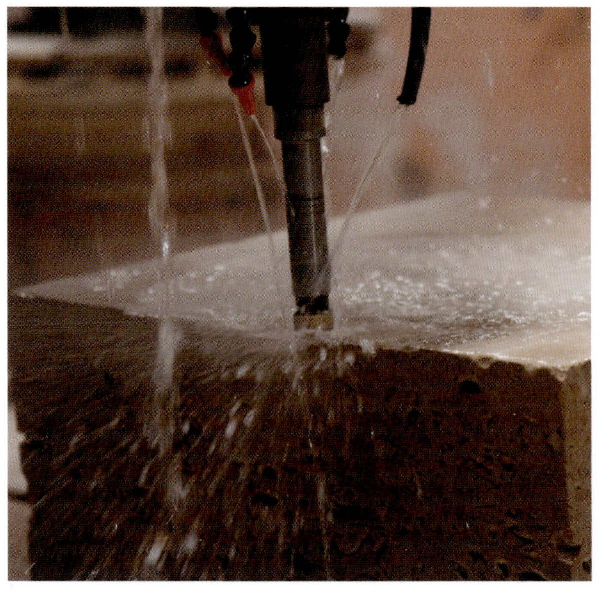

**PREVIOUS PAGES,
ABOVE AND RIGHT**

Null Object sculpture—
work in progress
2012

ARTISTS' INTRODUCTION

London Fieldworks

The *Null Object* sculpture has been manufactured from Portland Roach, a block of stone quarried from a limestone deposit laid down some 145 million years ago in the Jurassic geological period. The stone is highly fossilised with shells including Turreted Gastropods (Portland Screws) as well as large clams; an open textured oolitic limestone formed from an accretion of ooliths, sometimes known as ooids. The formation of ooliths started with a small fragment of sediment acting as a 'seed', e.g. a piece of a shell. Strong intertidal currents washed the 'seeds' around on the seabed, where they accumulated layers of precipitated calcite. Over time, countless billions of these ooids or ooliths became partially cemented together by more calcite to form the oolitic limestone we now call Portland Stone. The Turreted Gastropods and clam shells entombed within the limestone are associated with voids formed through the removal of the fossil shells by percolating rain.

The block of stone has been cut into a perfect 0.5 m cube, and has been largely excavated through the actions of a KUKA industrial robot; it has been rendered hollow as a consequence of the artist Gustav Metzger thinking about nothing.

So, in effect the *Null Object* sculpture is a cube of Jurassic stone peppered with tiny voids created through natural processes over multiple millennia; encompassing a large void within—the unthunk—created by a complex technology-art process executed at a relatively lightning speed.

Null Object has been conceived by London Fieldworks (artists Bruce Gilchrist and Jo Joelson) and draws on Gustav Metzger's radical engagement with environmental destruction through his art and political activism to inform a poetic application of technology; connecting a brain-machine interface, computers and bespoke software to an industrial manufacturing robot in the production of plastic art functioning through complex layers of body, installation, biofeedback technology, physiological data, computer database and software.[1]

In the summer of 2011 we hosted a research meeting between the artist and writer Armin Medosch and Gustav Metzger at our space in Hackney. Bronac Ferran, the instigator of the meeting, was also present. Afterwards the conversation turned to plans around the re-imagining of Event One, the first exhibition by the Computer Arts Society (CAS) in 1969 held at the Royal College of Art in London. The membership of CAS has proved to be remarkably prescient in recognising the long-term impact that the computer would have on society, and it is highly significant that Metzger, as a representative of the avant-garde was the inaugural editor of *PAGE*, the CAS journal. It was after this meeting that Joelson suggested inviting Metzger to participate in a project with London Fieldworks. We proposed connecting him via the 'Perception Depository Database' to some kind of manufacturing technology to function as a neurophysiological trigger in the production of an artwork.

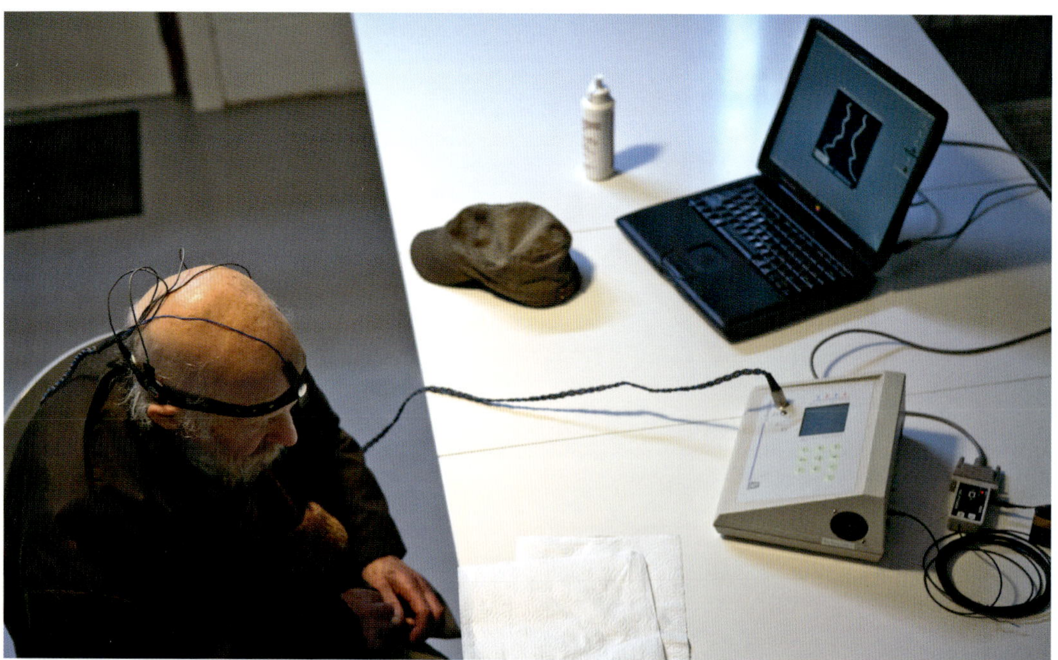

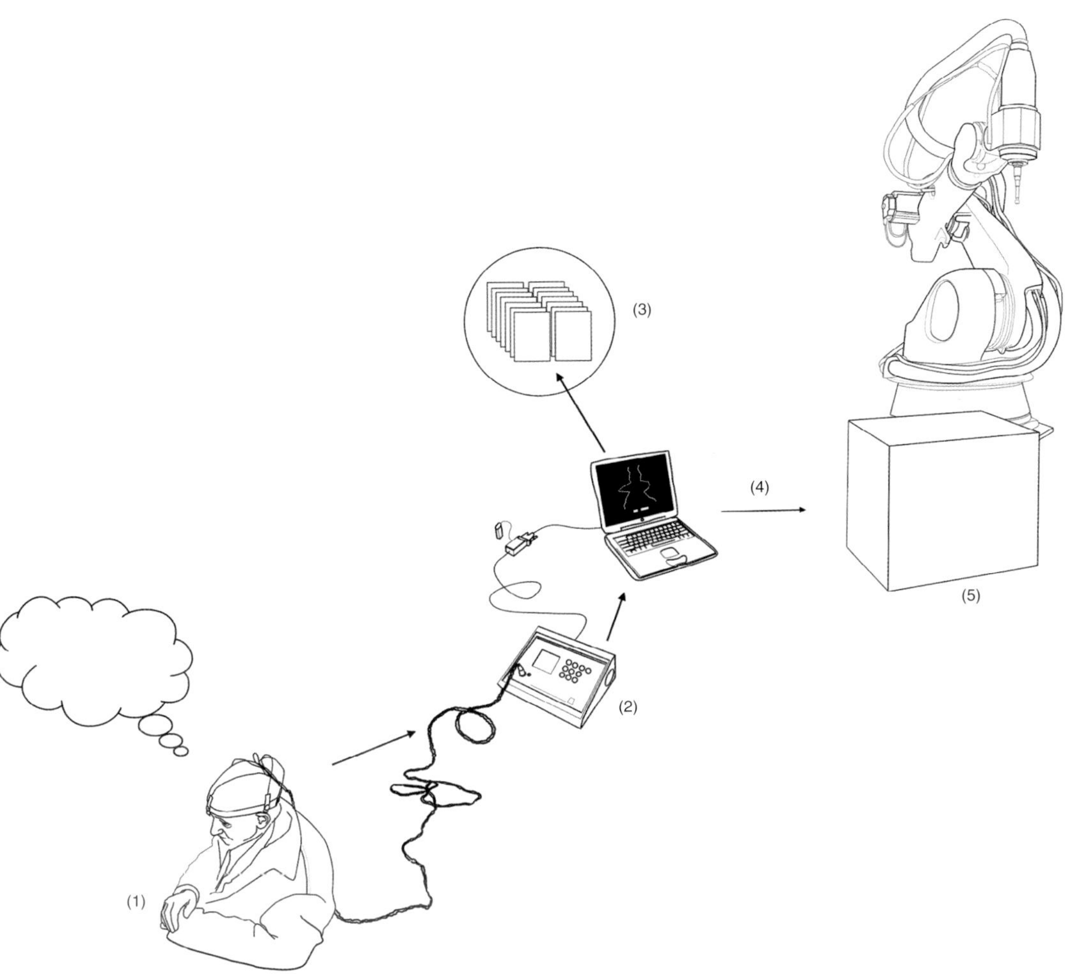

(1) The artist thinks about nothing. (2) The artist's brainwaves as a consequence of thinking about nothing are monitored by a digital power spectrum EEG machine. (3) Software analyses the brainwaves using a relational database to produce 3D shape information. (4) The 3D shape information is converted into control instructions for a manufacturing robot. (5) The robot excavates the shapes out of a block of stone to create a void.

The idea to 'think about nothing' as an instruction to activate the process of production previously came from a synthesis between the phenomenology derived from perceiving depth within random-dot autostereograms, and the Chilean biologist and neuroscientist, Francisco J Varela's notion of "The Portable Laboratory".

Null Object
process schematic
2012

Human beings in their embedded, situated life, de facto constitute a topographical place (the body, the self) where procedures and gestures can be carried to directly explore human experience itself (the quest). As in other laboratoria, the procedures present: follow shape and bring forth the content of what can and will become manifest. In the traditions of human wisdom (most notably Buddhism, Hinduism and Taoism) this portable, self-laboratory is the place for human discovery and transformation.[2]

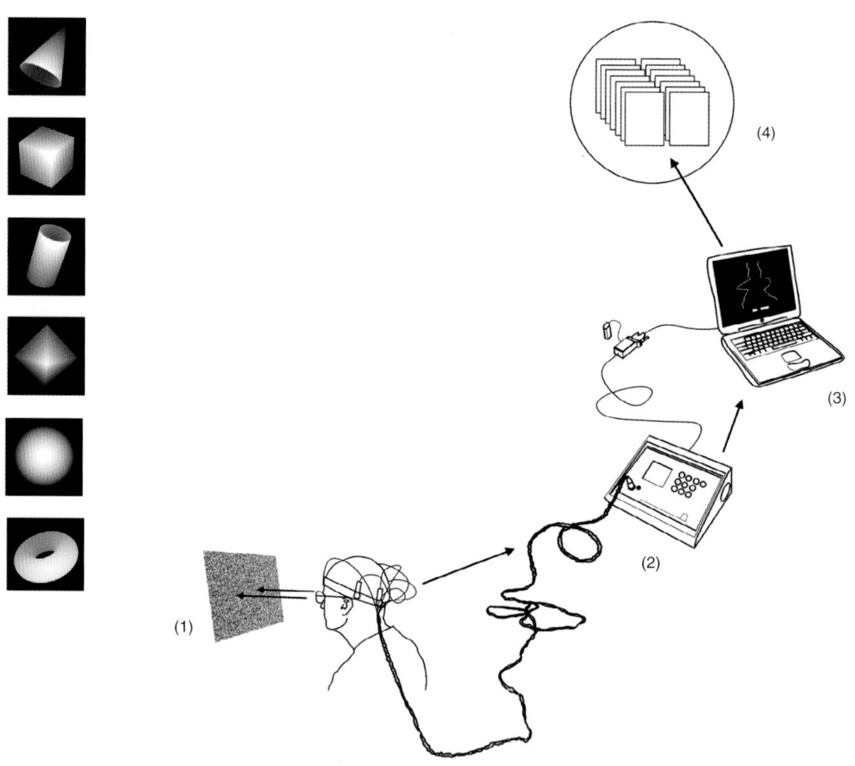

(1) Subject perceives depth-shape within random-dot autostereogram. (2) Subject's brainwaves are monitored using a digital power spectrum EEG machine. (3) Brainwave file is recorded during the act of perception. (4) Brainwave file and associated depth-shape is stored in a relational database: The Perception Depository.

Looking at Primitives and
The Perception Depository

Since 1999, the *Looking at Primitives* project has generated the database used in *Null Object*.[3] [see image opposite] The word 'primitive' used here in the project title refers to the generic primitive shapes—virtual building blocks—within the CAD software tools that design our contemporary environments. The project database, dubbed 'The Perception Depository', is partially derived from generic primitive shapes used in the construction of a series of random-dot autostereograms. The database contains participants' EEG files recorded whilst perceiving depth within the autostereograms, alongside the particular generic primitives that were perceived. *Looking at Primitives* draws on the field of psychophysics and the work of Béla Julesz and Christopher Tyler.

Psychophysics is a subdiscipline of psychology dealing with the relationship between physical stimuli and their perception. The random-dot autostereogram has evolved from a psychophysical instrument within the toolbox of cognitive science: the random-dot stereogram, developed by the visual neuroscientist and experimental psychologist, Béla Julesz at Bell Laboratories in 1959. In his book, *The Foundations of Cyclopean Perception*, Julesz explains how this imagery was created to investigate the neurological basis of depth perception: how cortical processes fuse the binocular 2D information from the optical nerves to create our experience of a 3D world.

The mythical Cyclops looked out onto the world through a single eye in the middle of his forehead. We too, in a sense, perceive the world with a single eye in the middle of the head. But our cyclopean eye sits not in the forehead, but rather some distance behind it in the areas of the brain that are devoted to visual perception. One can even specify a certain site in the visual system as being the location of the cyclopean eye. For instance, we can locate the cyclopean eye at a place where the views of the two external eyes are combined.... During the last decade, however, the cyclops within us has begun to collaborate with the psychologist and the digital computer. The result of this collaboration has been a series of unexpected insights into the nature of visual perception.[4]

Christopher Tyler (Head of Brain Imaging at Smith Kettlewell Eye Research Institute (SKERI) in San Francisco) worked with Julesz as a research fellow at Bell Labs. While at SKERI in 1979, Tyler significantly developed Julesz's random-dot stereogram—a stereo pair of images—by inventing the first single-image random-dot stereogram or random-dot autostereogram, with an Apple II computer connected to a dot-matrix printer. This invention made it possible for a person to see 3D shapes from a single 2D image, without the aid of a stereoscope or anaglyph glasses. Tyler's invention came to the global public's attention when popularised in a series of books published by Magic Eye (formerly N.E.Thing Enterprises). In order to perceive 3D shapes in these autostereograms one must overcome the normally automatic coordination between focus and vergence, or angle of one's eyes.

Looking at Primitives and the development of the database took place in a variety of venues from 1999–2012, including Artec, Strike studios, the Foundry, the Toy Factory, London Fieldworks' studio in Hackney and the V&A Museum in London; Farnham University, Oxford Brookes University, UK; the Oslo Contemporary Art Museum, Norway; Center for Consciousness Studies, Tuscon, Arizona, the LAB gallery, San Francisco, USA.

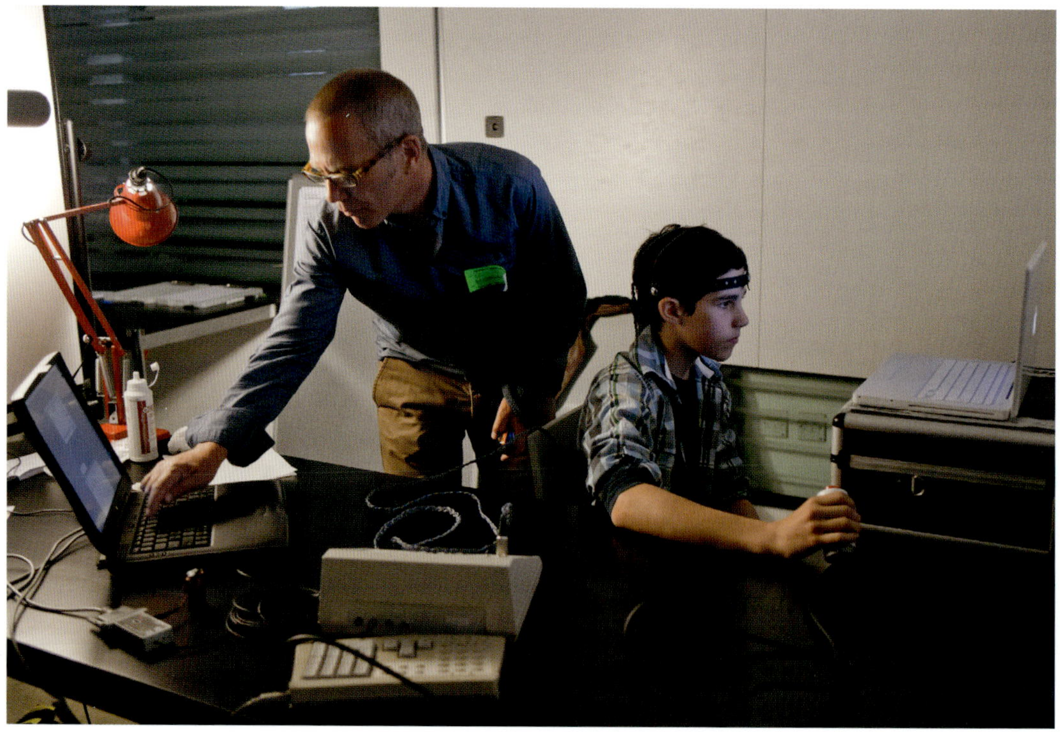

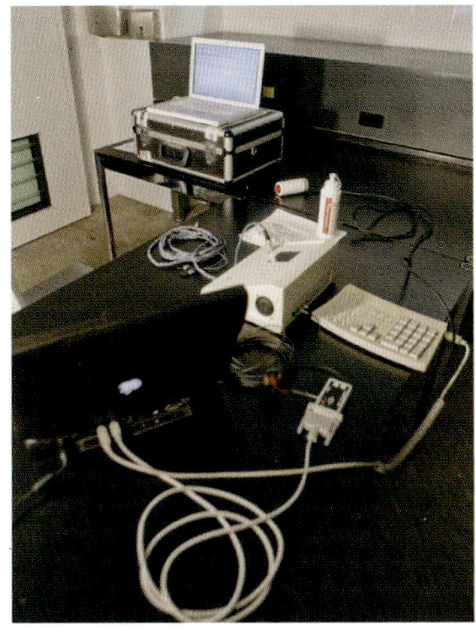

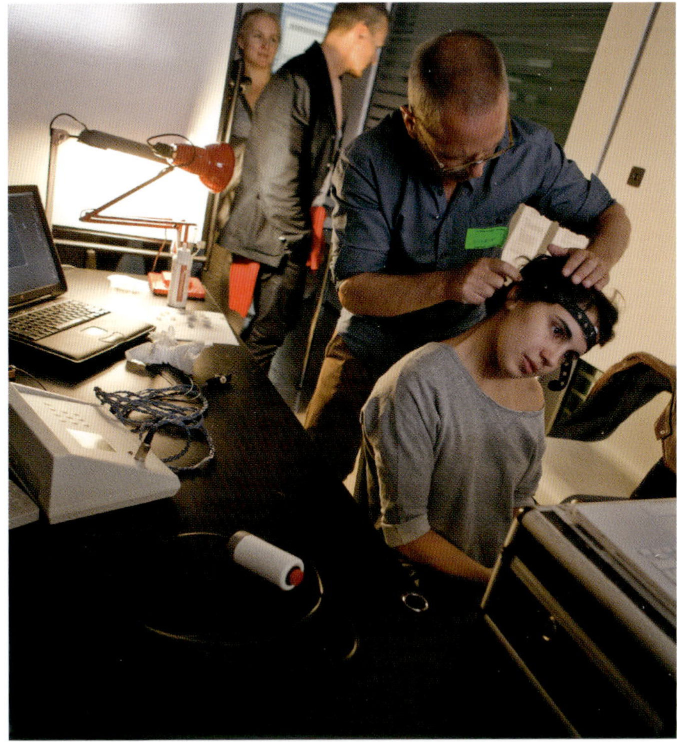

Looking at Primitives
Victoria and Albert Museum,
London, 2012

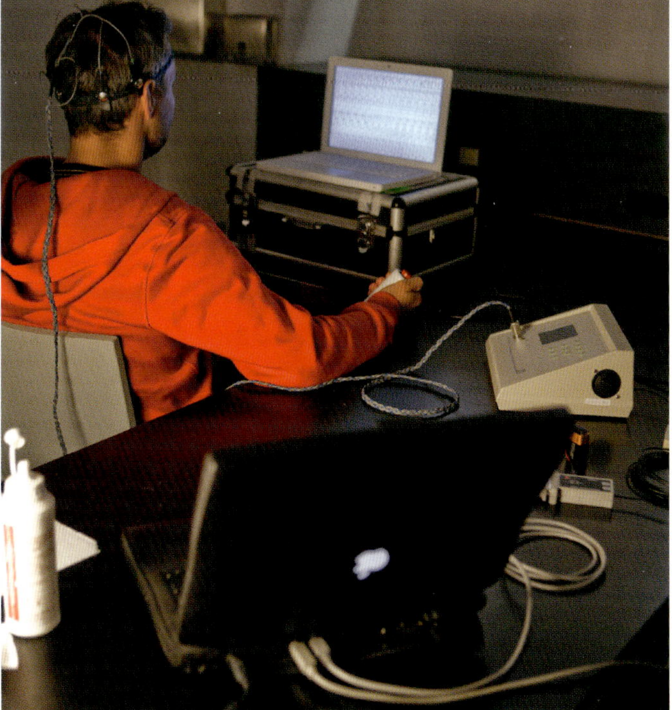

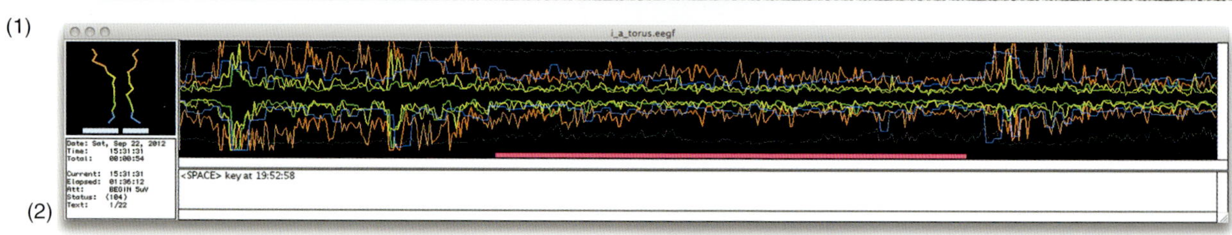

(1)

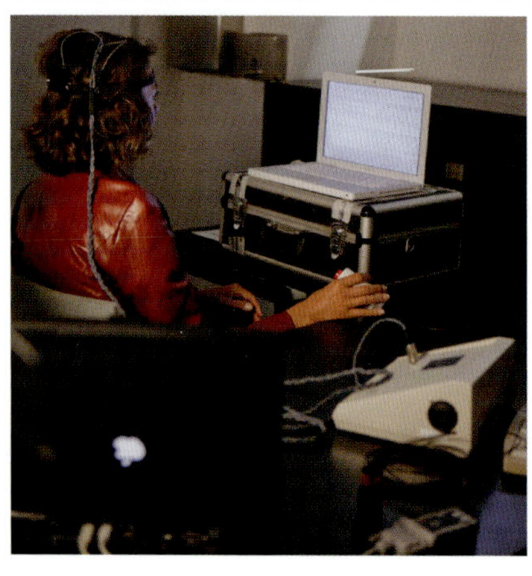

(2)

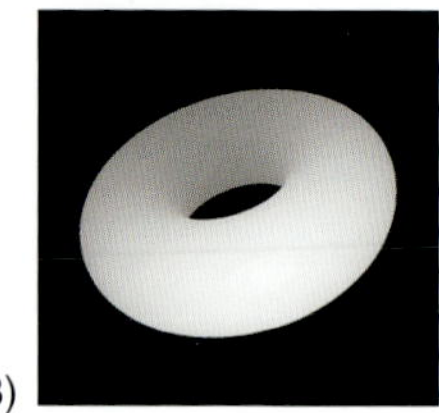

(3)

THIS PAGE
EEG file recorded
at Digital Weekend
Victoria and Albert
Museum, London, 2012

OPPOSITE PAGE
EEG file recorded
at Strike Studios,
London, 1999

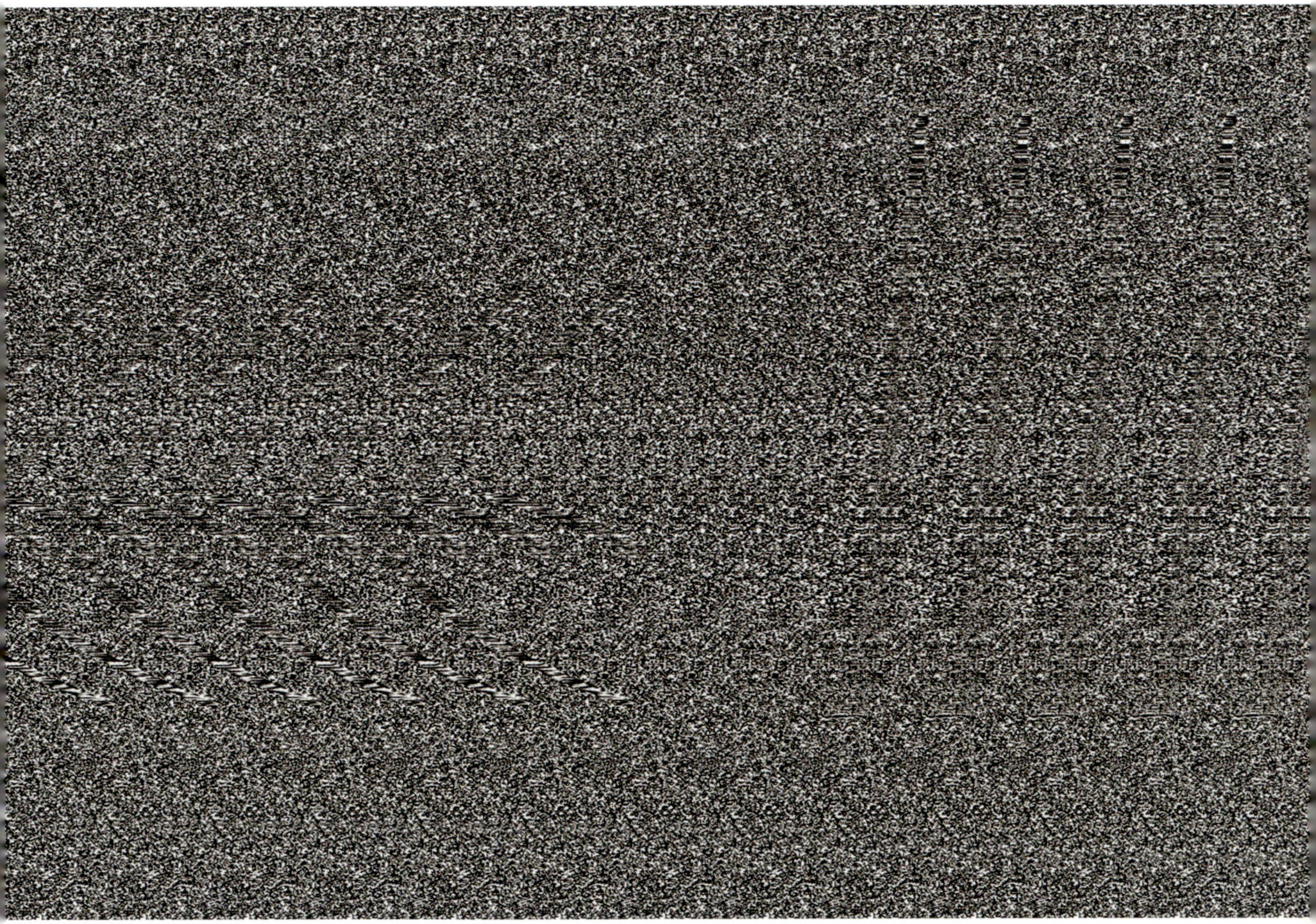

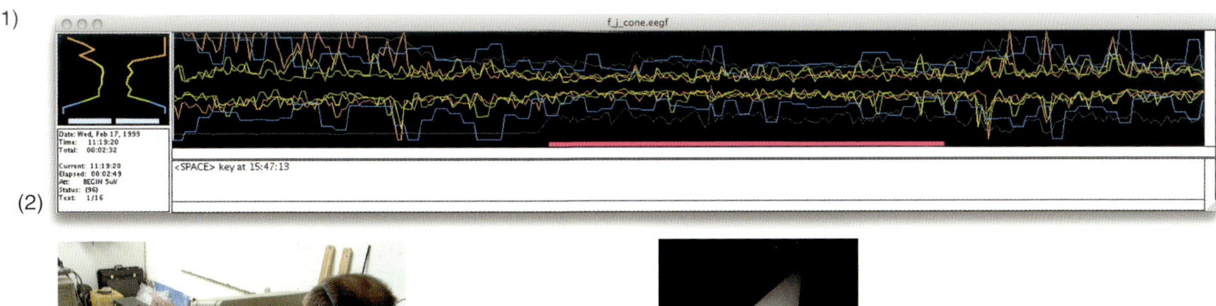

(1)

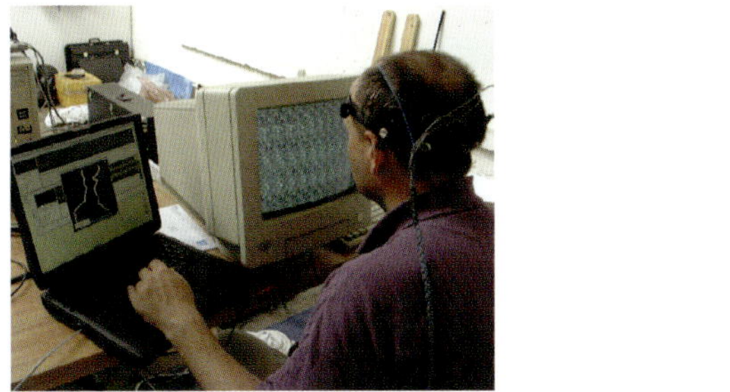

(2)

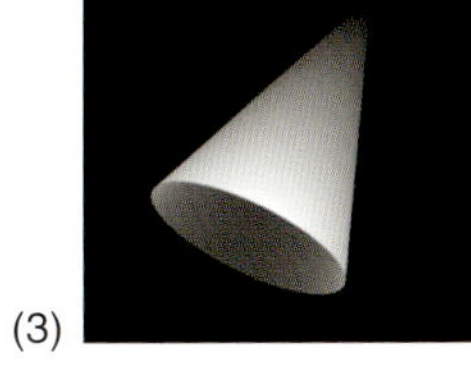

(3)

The database contains brainwave files from members of the public perceiving a series of random-dot autostereograms. Each brainwave file in the database has the primitive shape perceived at the time of recording associated with it. The autostereogram (1) has an embedded primitive shape (3) as its depth-image. The brainwave file (2) was recorded while a subject perceived the shape within the autostereogram. The red line marked on the brainwave file indicates the time and duration the subject perceived the depth-shape.

(1)

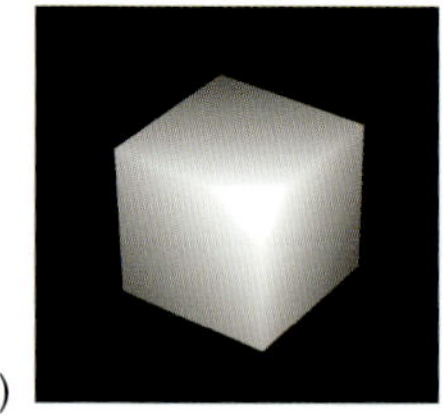

(2)

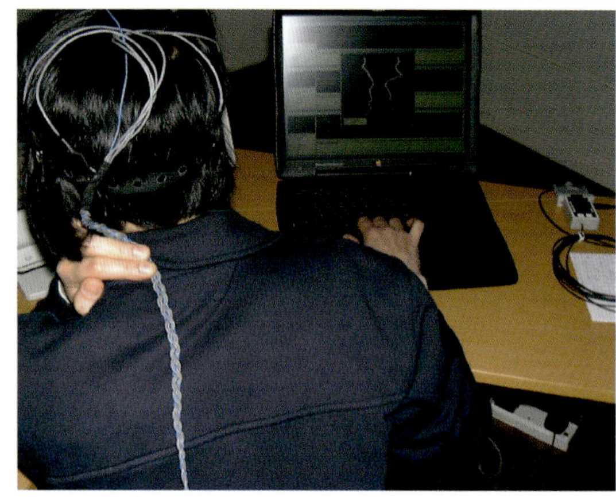

(3)

THIS PAGE
EEG file recorded at
Oslo Contemporary
Art Museum,
Norway, 1999

OPPOSITE PAGE
EEG file recorded at
the LAB Gallery,
San Francisco, 2002

(1)

(2)

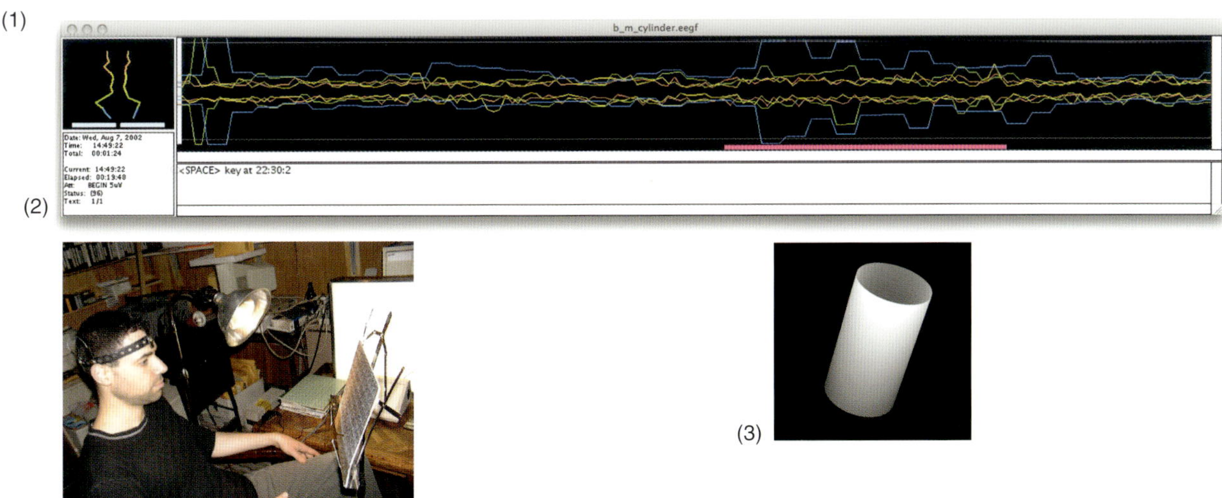

(3)

The database contains brainwave files from members of the public perceiving a series of random-dot autostereograms. Each brainwave file in the database has the primitive shape perceived at the time of recording associated with it. The autostereogram (1) has an embedded primitive shape (3) as its depth-image. The brainwave file (2) was recorded while a subject perceived the shape within the autostereogram. The red line marked on the brainwave file indicates the time and duration the subject perceived the depth-shape.

From its earliest iterations this larger project is resonant with the "vision machines" described by Paul Virilio: the "industrialisation of vision" and "mechanisation of perception" through a desire to mass communicate: who chooses what we see and how we see it, and how this affects our thinking about the world, notions of society and the solitary self.

> The look of direct vision is thus ceding to a real-time, radio-electrical industrialisation of vision capable one day of standing in for, if not supplanting, our observation of the environment. The direct light of the sun, candles or electric lamps will gradually make way for the not just artificial but indirect light of electronics or photonics, following the example of those windowless Japanese apartments that are bathed in sunshine by means of optical fibres.[5]

The neurobiologist William Calvin recognises that the science and technology of mind may move far more quickly then we can create consensus for; creating another level of stratification between "The Enhanced and The Rest".[6] Our project engages with the idea of augmented mind, by responding to the development of brain-machine interfaces, biometrics and the burgeoning role of databases across all conceivable sectors of society, from the commercial sector to government and law enforcement.

> Our personal human data is interpreted and employed by remote, inhuman IT and statistical interfaces.... Commercial exploitation of our data attempts to corner us in an inescapable labyrinth of consumption... the twenty-first century body is enmeshed in contradictions of fixity and flux, authenticity and simulation.[7]

Thought Pavilion

A Science Museum Art Projects 5 (SMAP5), Big Ideas commission in 2005, led to a further conceptual development to integrate architectural principles into the project software to generate a building—a "Thought Pavilion". It was proposed that 3D objects could be produced via a plaster-based 3D printing technology as building components; each object constituting a building block within a larger accumulated architecture that could be inhabited and used by members of the public. During this development period an

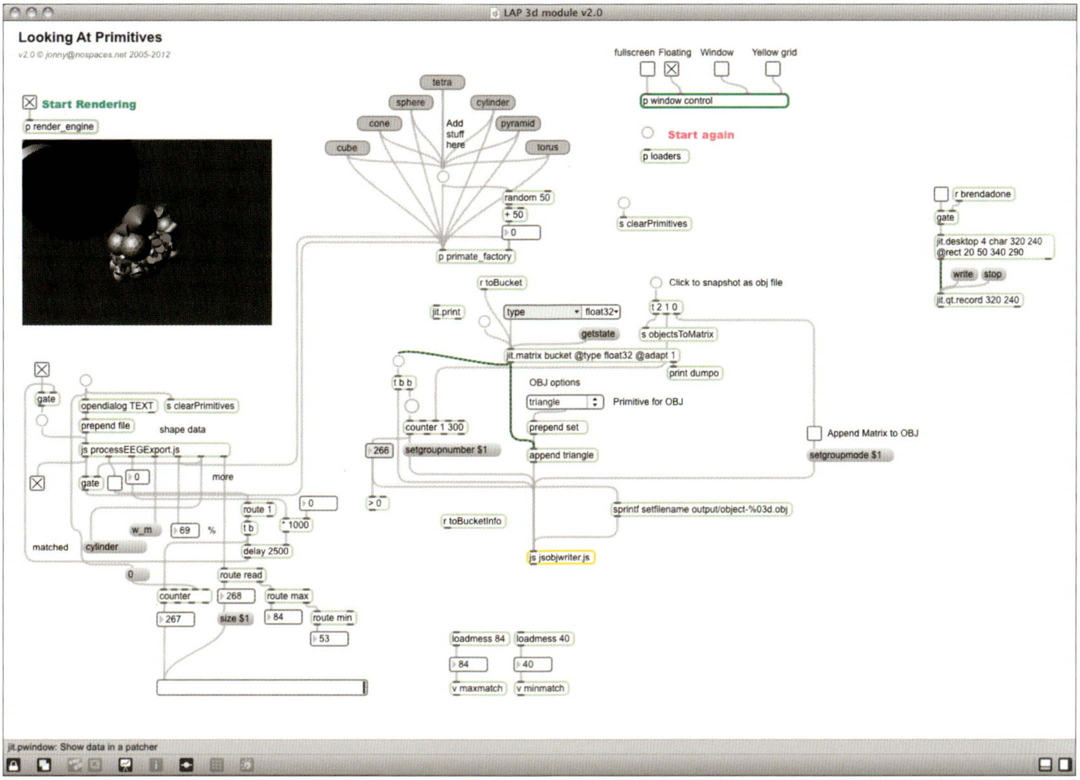

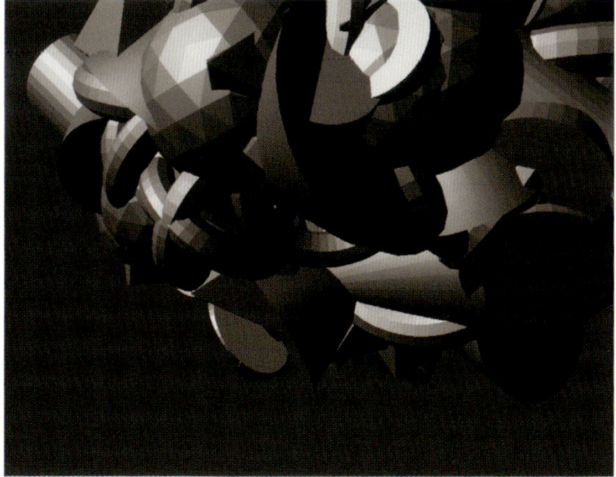

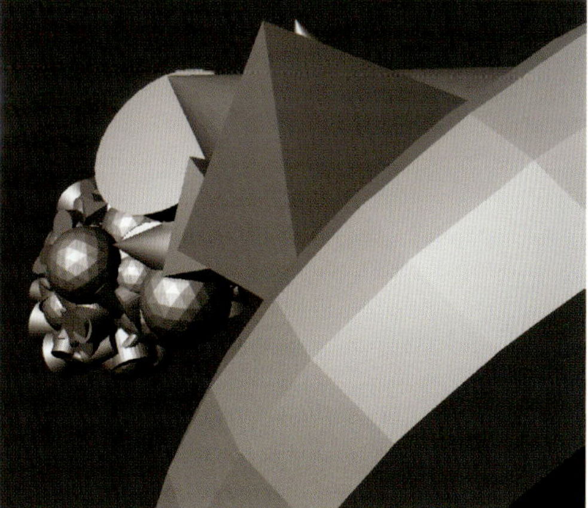

Null Object: Looking at Primitives
3D Module software
2012
Images courtesy Jonny Bradley

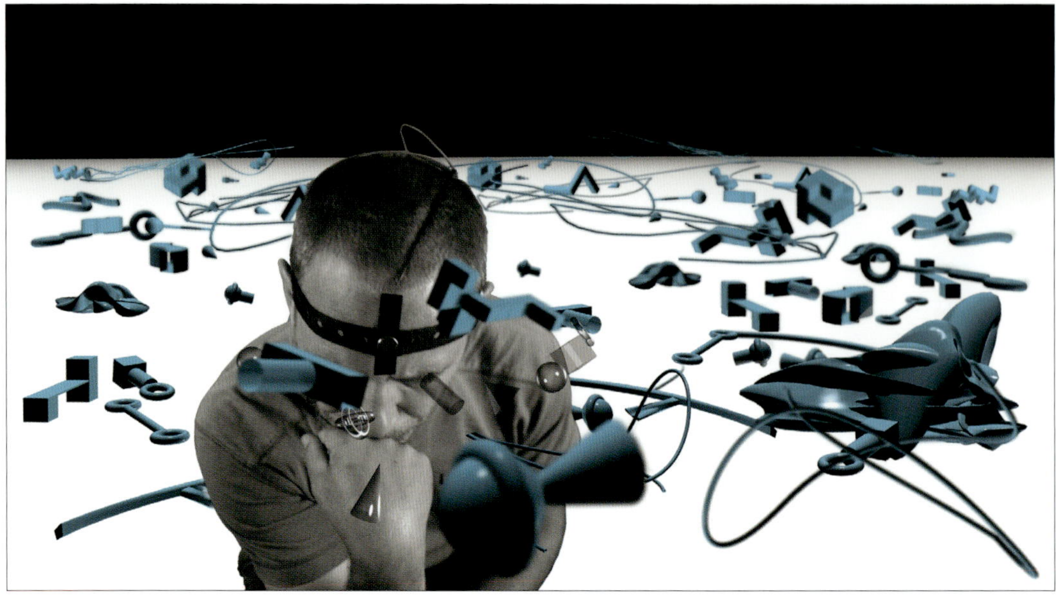

ABOVE

Thought Pavilion
Digital collage, 2005

OPPOSITE TOP

Divided by Resistance
ICA, London, curated
as part of Totally Wired, 1996
Photography Bonnie Venture

OPPOSITE BOTTOM

Divided by Resistance
ICA, London, curated
as part of Totally Wired, 1996
Video stills

exploratory meeting was held with parametric modelling experts at the HQ of Arup in London. The project was approved by the Ethics Committee at London South Bank University, however to date *Thought Pavilion* has not been realised.

> Can a participant [in *Thought Pavilion*] claim intellectual copyright to brainwaves captured and any resulting 'artworks'? In the light of the ethical and moral issues surrounding the Human Genome Project, could the potential creative application of the contents of The Perception Depository be considered a public work of art or an artwork disinterred from the public?[8]

Developing The Database Aesthetic

In 1993, Bruce Gilchrist visited the Clinical Psychology Dept at The University of Texas at Austin where he was introduced to the methodology of the sleep lab and the principles of biofeedback. That research visit catalysed a sustained investigation that combined poetic strategy with customised brain imaging and biofeedback technology to elicit expression from mute bodies in variegated states of consciousness. His artworks "journey in to the unconscious, unlanguaged parts of consciousness, into prehension and preconsciousness" in an exploration of "what is in between the languaged and the super-ephemerality of consciousness".[9]

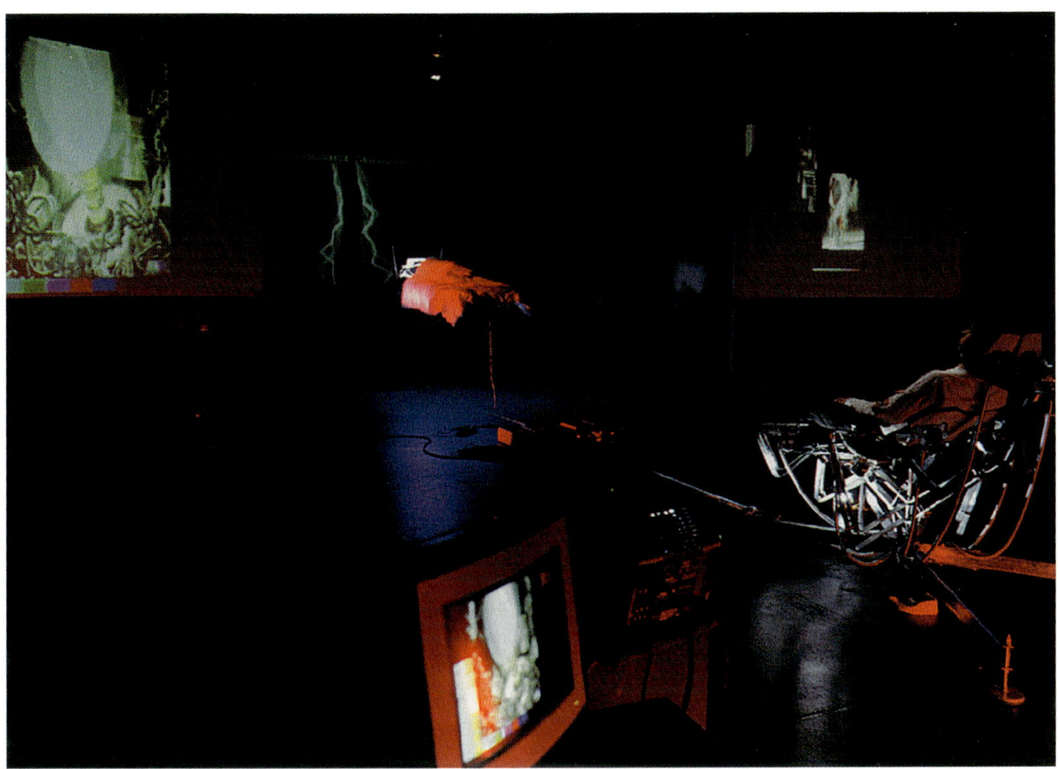

Divided by Resistance

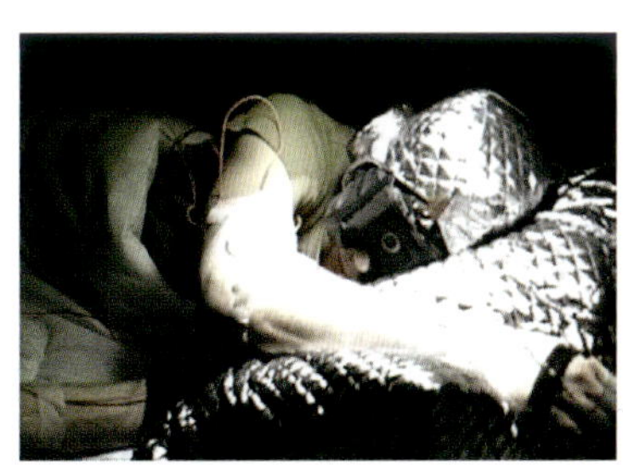

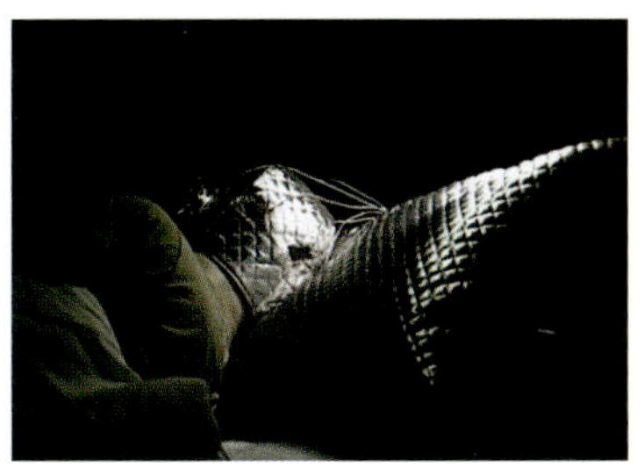

At the core of the *Null Object* project is a database aesthetic: creative records are retrieved and presented in the manner of a search engine.[10] The software used in this project has largely evolved from an ICA/Toshiba Art & Innovation commission awarded to Bruce Gilchrist and programmer Jonny Bradley for the digital live-art installation, *Divided by Resistance*, presented by the ICA in London as part of Totally Wired in May 1996.

Divided by Resistance utilised a relational database of EEG (electroencephalograph) recordings, Quicktime movies and text narratives based on recollections from periods of REM (rapid eye movement) sleep. This data was collected intensively over a period of one year, from the artist sleeping in his studio. For the final performance at the ICA, for which Gilchrist had inverted his sleeping patterns several weeks before, members of the audience were invited to interact with him via a technological communication conduit, eliciting responses from the sleeping artist in the form of re-mixed Quicktime video clips and cut-up texts.

"The title of *Divided by Resistance* suggests the solipsism of the individual never able to fully communicate across the divide from one subjectivity to another."[11]

> Although developed from a research context that is generated by sophisticated scientific concepts, Gilchrist's performances also function on another, less immediately apparent level which is dependent upon establishing a sensitive physical dialogue between artist and individual audience member.... We are aware of altered states of sleep, of the possibility of uncovering despair in the subconscious. We are uncertain about making the journey into the scientifically unknown. It is hard to find the subtle channels of access. There are two possible responses to this. We may feel alienated by the technology and daunted by the possibility of crisis which we refuse to acknowledge, or we may on the other hand see in the work a reflection of our own impossibility and in so doing understand something of what is meant by 'making strange' and move a little way towards a perception of what was hitherto unknown or unseen.[12]

Thought Conductor

The original software written for *Divided by Resistance* was subsequently developed for the series of *Thought Conductor* performances, inspired by the notion: what could have occupied the mind of the musician David Tudor when he performed John Cage's silent composition 4'33"?[13] In reality he might have been preoccupied with keeping track of time, but this led to the idea of a musical performance where musicians are responding not to an arranged score, but instead to a direct manifestation of a conductor's thought processes on stage. For the *Thought Conductor* performances, with the aid of a biomonitor, passages of musical notation re-mixed from the contents of a database were called up in the real-time of the performance.

> ... score and sound acquire an immediacy which is characteristic of neuroscientific imaging in general, but which in this specific context lends new meaning to the notion of a 'live' performance. It is tempting therefore to suggest that *Thought Conductor* is a techno-scientific portrait that captures the inner kinetic melody of the individual who sits at its centre....[14]

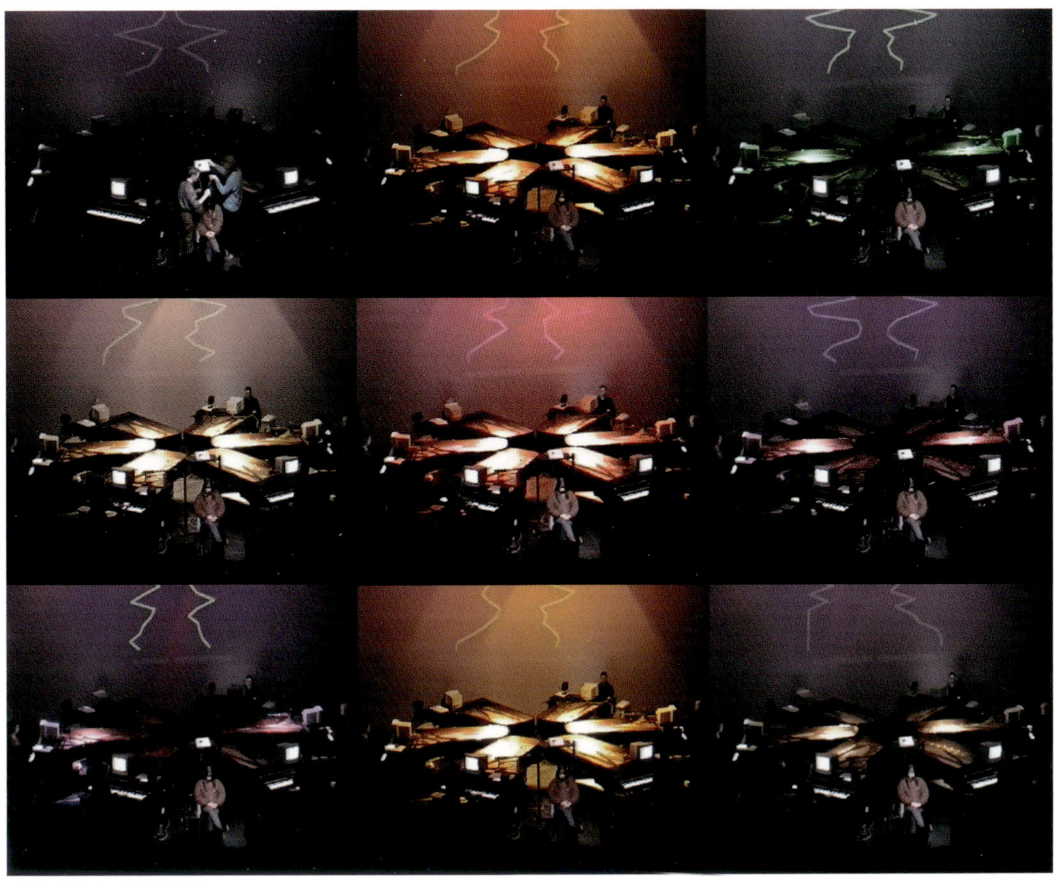

Thought Conductor #1
Wintermusic Festival
The Place, London, 1997
Video stills

#1

For the Wintermusic Festival in 1997 the composer, Nikki Yeoh was commissioned by the piano sextet, Piano Circus to compose a new piece for the festival that she titled, *"six-as-one"*. She consented to a *Thought Conductor* collaboration. Software was written that would translate the composer's real-time brainwaves into musical notation based on mathematical rules agreed between the programmer and the composer. Prior to the world premiere of the authentic *"six-as-one"*, the composer was connected to a digital biomonitor on stage and asked to think through her score, adhering as much as possible to the structure of the piece. Her real-time brainwaves, as a consequence of this act, were translated into musical notation for six parts and sent to six computer monitors, each one mounted on a Steinway grand piano. The six members of Piano Circus sight read the scrolling notation and played it live on stage.

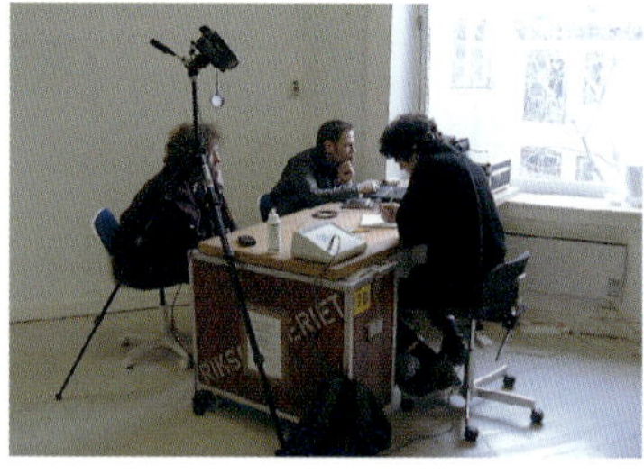

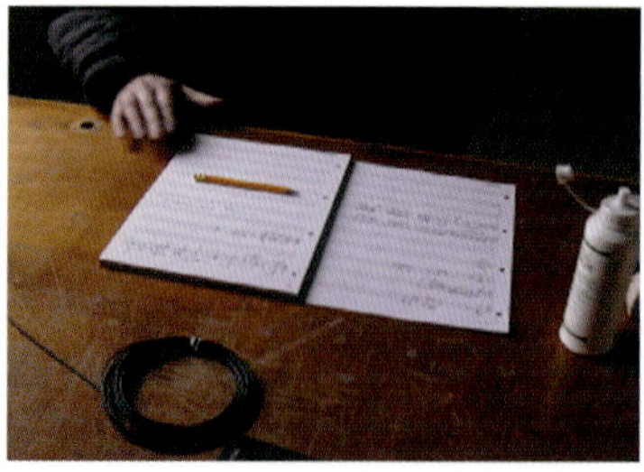

RIGHT

Thought Conductor #2
Recording composers' EEG
in the act of composing music
for a string quartet
Studio Struts, Oslo, 2000

#2

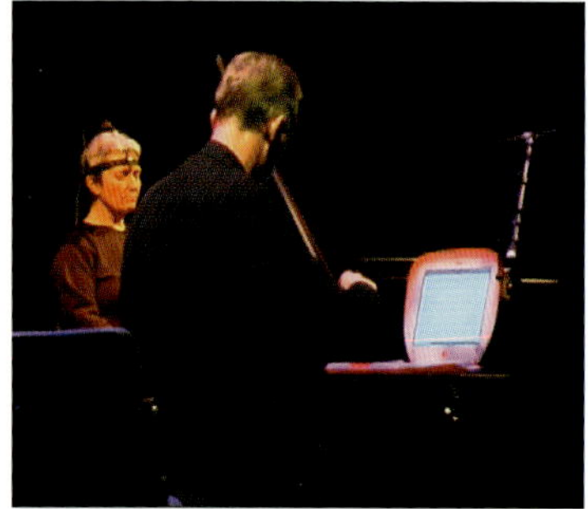

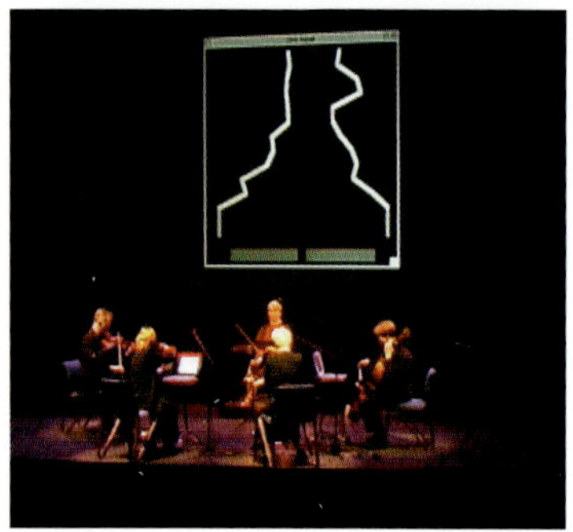

ABOVE

Thought Conductor #2
Speculum Combi
IBCA 2000 Festival
Stavanger, Norway, 2000
Video stills courtesy
Svein Flygari Johansen

An invitation from the Speculum Combi IBCA 2000 Festival in Norway provided the impetus to further develop the *Thought Conductor* concept, this time integrating the database principle originally conceived for *Divided by Resistance*. Several weeks before the performance at the Stavanger Konserthus, Gilchrist and Bradley were resident at Studio Struts in Oslo, to collaborate with a number of Oslo-based composers.

12 composers in all were invited to the studio and asked to write spontaneous compositions for a string quartet. While writing the musical notation, their EEG was recorded. At the end of this process, the recorded EEG file was associated with a midi version of their notation and archived within a database. For the Stavanger Konserthus performance four members of the community were invited on stage in succession and connected to the biomonitor. They were then asked to think of their last creative act. In the event, a bricklayer mentally recollected building a brick wall; a jeweller thought about a gold ring he had recently made; a yogi meditated; a painter imagined the process of making her last painting. Each 'thought conductor' generated original scores from the notation within the relational database, each part of which was sent to laptops for the string quartet to sight read and play live on stage.

#2.1

For the OX1 Festival in 2001, the database archive made from the Oslo based composers was supplemented with EEG files and notations from six Oxford-based composers. The same procedures were followed. For the performance at the Holywell Music Rooms, Oxford, several members of the festival audience participated. Again they were invited on stage, connected to the biomonitor and asked to engage in creative thought: an origami enthusiast recollected making a paper lotus flower; an accountant thought about a financial projection for a board of directors; a journalist recalled her last piece of arts criticism; an academic contemplated his tie.

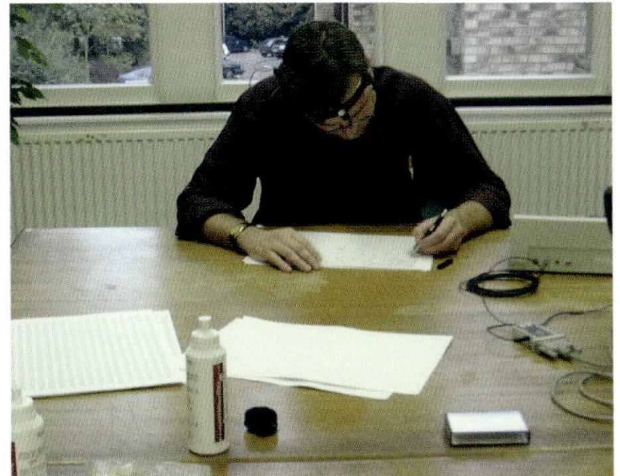 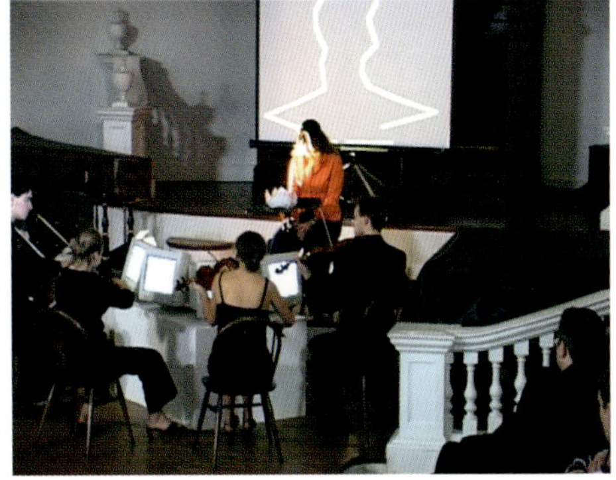

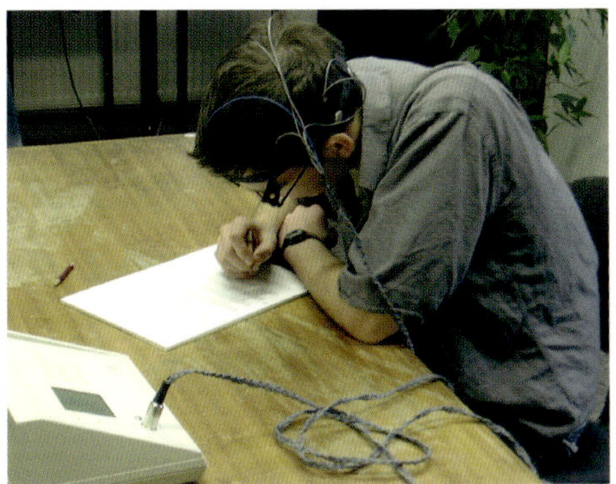 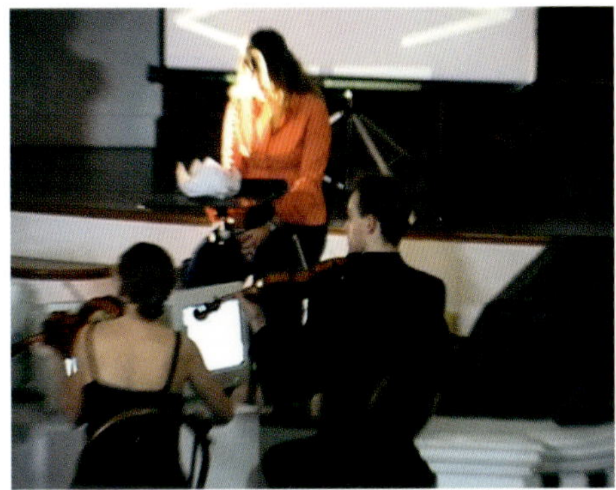

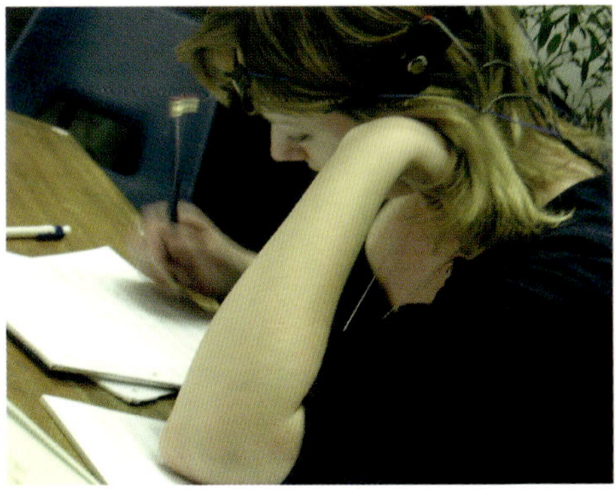 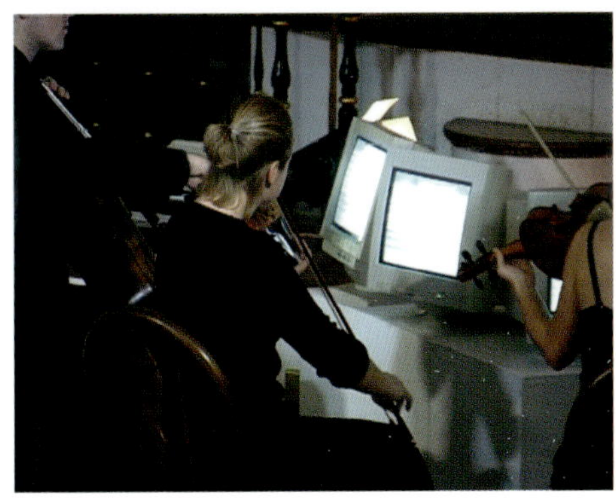

LEFT
Thought Conductor #2.1
Recording composers' EEG
in the act of composing music
for a string quartet
Oxford Brookes University, 2001

RIGHT
Thought Conductor #2.1
OX1 Festival
Holywell Music Rooms,
Oxford, 2001

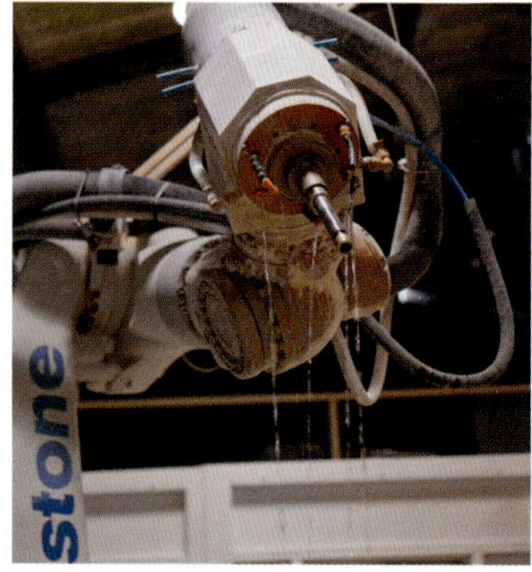

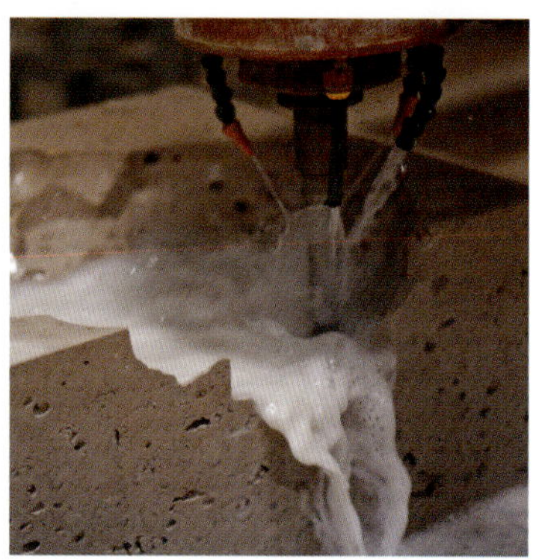

Null Object sculpture—
work in progress
2012

As a point of departure from the works previously described, which have variously rendered evanescence in temporary audio-visual forms, it had been imagined that *Thought Pavilion* would employ an additive manufacturing process such as stereolithography or 3D printing, a method and apparatus for making solid objects by printing successive layers of material one on top of the other. With the participation of Gustav Metzger, there was a shift of context, allowing for the imagining of a subtractive process, an excavation or voidance, connecting the concepts of the threshold of thought and the removal of material. The evanescent faded into the unthought, not as something external to thought but something at the very heart of thinking.

In her book, *Symbol and Myth*, Barbara Maria Stafford refers back to Enlightenment and Romantic enquiries into the beginning of thought, revealing something vital about how the brain generates its sense of reality, something which certain contemporary neuroscientists are struggling to express again—namely, that "the other world is always also the inner world".[15] After having linked neurological research on various altered states to rock art in her chapter, "Primal Visions", Palaeolithic caves are presented as echoic images of the cave in the mind—the depths of the unconsciousness— where, if one cares to look, "the secret of self-organisation embodied in primal images of the simplest forms will be found hiding".[16]

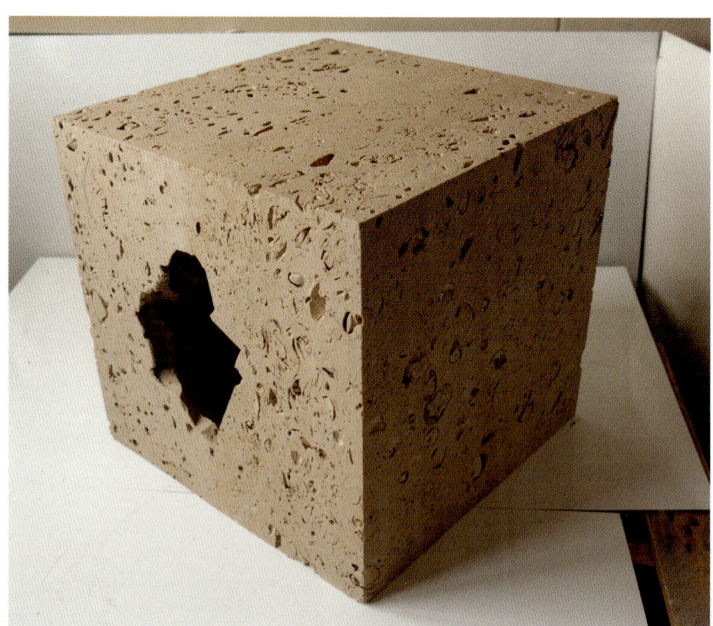

NOTES

[1] Plastic art: an art, as painting or sculpture in which forms are carved or modelled; are rendered in or as if in three dimensions.

[2] *Laboratorium*, (exhibition catalogue), Antwerp: Antwerpen Open, 1999. Curators: Hans Ulrich Obrist, Barbara Vanderlinden.

[3] Supported by an AHRC research fellowship at Oxford Brookes University, 2002–2005.

[4] Julesz, Béla, *Foundations of Cyclopean Perception*, Chicago: The University of Chicago Press, 1971, p. xi.

[5] Virilio, Paul, *Polar Inertia*, London: Sage Publications Ltd, 2000, p. 57.

[6] Calvin, Wiliam H, *A Brief History of the Mind*, Oxford: Oxford University Press, 2004, p. 172.

[7] Warr, Tracey, "Texts from the Body", *Sensualities/ Textualities and Technologies: Writings of the Body in 21st-Century Performance Art*, Susan Broadhurst and Machon, Josephine eds., New York: Palgrave Macmillan, 2009, p. 29.

[8] Gilchrist, Bruce, Tracey Warr, 2000, "Art as a First-Person Methodology in Consciousness Research", unpublished conference paper presented at Toward a Science of Consciousness at the University of Arizona in Tuscon in April 2000. Abstract published in *Toward a Science of Consciousness: Tuscon 2000*, Thorverton: Journal of Consciousness Studies, p. 162s.

[9] Warr, Tracey, "Texts from the Body", *Sensualities/ Textualities and Technologies: Writings of the Body in 21st-Century Performance Art*, p. 23.

[10] Manovich, Lev, Andreas Kratky, "Cinema and Software", *SOFT CINEMA: navigating the database*, Cambridge, MA: MIT Press, 2004, n.p..

[11] Walwin, Jeni, "Subtle Bodies: The Work of Bruce Gilchrist", *Low Tide: Writings on Artists' Collaborations*, London: Black Dog Publishing, 1997, p. 24.

[12] Warr, Tracey, "Texts from the Body", *Sensualities/ Textualities and Technologies: Writings of the Body in 21st-Century Performance Art*, p. 27.

[13] *Thought Conductor* performances: #1, Wintermusic, The Place, London, 1997; #2, Speculum 2000, Stavanger Konserthus, Stavanger, Norway, 2000; #2.1, Hollywell Music Rooms, OX1 Festival, Oxford, 2001.

[14] "Making Music Matter" by Mariam Fraser, Goldsmith's College, University of London, an academic article based on *Thought Conductor #2* can be accessed at tcs.sagepub.com/content/22/1/ 173.abstract

[15] Maria Stafford, Barbara, *Symbol and Myth: Humbert De Superville's Essay on Absolute Signs in Art*, Cranbury, NJ: Associated University Presses, 1979, p. 25.

[16] Hadravová, Tereza, "Barbara Maria Stafford: Echo Objects. The Cognitive Work of Images", 2008, available at www.researchgate.net/ publication/31913098_Barbara_Maria_Stafford_ Echo_Objects._The_Cognitive_Work_of_Images> [Accessed 7 October 2012]

SCULPTING THE VOID: BETWEEN NEGATION AND ABNEGATION, BETWEEN ACTION AND ABSENCE

Bronac Ferran

While you were watching, a rock was cut out, but not by human hands. It struck the statue on its feet of iron and clay and smashed them. Then the iron, the clay, the bronze, the silver and the gold were broken to pieces at the same time and became like chaff on a threshing floor in the summer. The wind swept them away without leaving a trace. But the rock that struck the statue became a huge mountain and filled the whole earth.

Old Testament[1]

It's odd that science, so they tell me, is making things (so to speak) more spiritual…. The very latest notion, so I am told, is nothing's solid.

Virginia Woolf[2]

Beware the tyranny of software!

Bruce Gilchrist[3]

Null Object combines hyper-performativity with auto-production of sculptural form. It fuses together 'mind-stuff' with software, wetware and hardware to produce a void in stone.

The work is the brainchild, literally and metaphorically, of Bruce Gilchrist and Jo Joelson working together as London Fieldworks. It has developed from a series of research experiments, action research and background thinking over many years. They have brought together an international, interdisciplinary team to produce a work which has involved hundreds of participants who have readily volunteered to take part in a perceptual art experiment.

The formative code that has driven the axis of connection between human, robot and stone has been written by Jonny Bradley, long-term collaborator with Gilchrist and Joelson. Medical roboticists Yaroslav Tenzer and David Noonan formerly of Imperial College and now both based in the USA (Tenzer at Harvard and Noonan at Philips in New York) have given expert advice with respect to finding the right robot and facilitating the transformation of 3D shape data into robot language, crucial for setting it to work to make the appropriate actions. The stereoscopic research of Béla Julesz, as described elsewhere in this publication, planted the original seed which inspired the direction of Bruce Gilchrist's investigations into identification of primitive shapes viewed within EEG (brain) experiments.

The work also draws politically and precisely on the mind of Gustav Metzger, artist and thinker and a close neighbour of Gilchrist and Joelson, living in Hackney. They first met during the production of *Gastarbyter* (from the German Gastarbeiter or 'guest worker', a reference to the German labour programme of the 1970s) devised by Gilchrist and Joelson for the ICA in 1998. In *The Independent* Judith Palmer wrote about the experience:

> Flowing red lines zip on and off, occasionally turning blue, as the neon hits a pocket of mercury. A low revving hum sends a brief tingle up one leg, a couple of deep bongs and your buttocks get a delicate pummelling, then, joy, a sustained vibro-massage of the lower back. It is like sitting watching the dying glow of a four-bar electric fire with each foot on the pedal of an electric sewing machine, reverberations travelling up and down the body. Lights and chair-sensors react in harmony with the shifting electronic noise, jiggle to a few short percussive tones, then surf down a long, mesmerising gong until you are literally inhabiting the interior landscape of the sound.[4]

Invited to enter a darkened theatre alone with no idea what would lie ahead, invited to lie back whilst sounds composed by Dugal McKinnon were streamed into ears and vibrations from the large steel tensile chair reverberated through bones, this was a work where absence of physical agency collided with psychic fear of the unknown.

A Zone of Deepest Silence

In a beautiful essay written about the work of Susan Aldworth, another Hackney-based artist making compelling visualisations of her own experience of consciousness, neurophysiologist Paul Broks has observed:

> Now here is the paradox: the zone of deepest silence is the brain. Three pounds or so of jellified fats, proteins and sugars packed into the bony box of the skull, numb and dark as a tomb. But look at the spirits summoned up from the sludge. Look at the carnival of consciousness that flows from that void. How is that possible?... This is the Century of Neuroscience. Images of the brain have become commonplace.... Modern neuroimaging methods—PET, fMRI, MEG—bring the metropolis to life. Trails of colour trace the circuitries of memory and emotion.... The harder one stares into the machinery of the brain, the starker the realisation that there is no one in there. There is no inner sanctum of the self. Neural networks have a life and logic of their own. There is no one running the show. The self is a shadow-puppet shaped by the firings of a hundred billion brain cells. These are conceptual conundrums, intractable to current science, they call for an artistic response.[5]

Thought Pavilion

After three years working on the foundational research for the work as part of a research fellowship at Oxford Brookes University, Gilchrist with Joelson, developed a proposal in 2005 for a work called *Thought Pavilion* submitted to an arts programme run by Hannah Redler at London's Science Museum. Whilst it proved difficult to secure full funding for this project, which also would have built on the contributions of volunteers, the process of dialogue was a useful one, for implicit even at this research and development stage were complex questions which are now far more prevalent within mainstream media discourse to do with control, access to private space, mutual responsibility in terms of sharing knowledge, boundaries of the so-called public domain, ownership of images and thoughts and manipulation of data for the purposes of exhibition and display. The ethical implications were such that it was necessary for the research to be approved by an ethics committee and with support from Arts Council England's Interdisciplinary Arts Department's art and law initiative, the appointment of art law advisor Sarah Andrew was made. Subsequently the Ethics Committee of London South Bank University approved the project.

As is characteristic of London Fieldworks' often anticipatory practice, *Thought Pavilion* raised issues, which are now regularly debated within zones of bio-art and medical humanities including often contested questions of donation by members of the public of bones, tissues etc. for exhibitions and collections or for purposes of artistic research. Whilst current discourse around this tends to be focussed on objects with material form the interesting thing about *Thought Pavilion*, and now taken forward with *Null Object*, is the dematerialised nature of what is being offered by volunteers within the project i.e. their capacity to see, or in the case of Metzger, to think. The questions arising from this are necessarily ethical and philosophical.

The Ethics of Machines

Null Object has also forced into the light vital intellectual and ethical questions which need to be thought about in terms of design and which are specifically raised by the project's mirroring of processes of automation, distributed and decentralised responsibility familiar within industrial design. As articulated clearly by Vilém Flusser, the twentieth century phenomenologist and philosopher, great risks exist in the gaps between thinking, making and executing:

> The only authority that seems to me more or less intact is science. Of course, it always claims to be engaged in value-free research, and as a result it does provide technical norms but not moral norms. Second, industrial production, including design, has developed into a complex network that makes use of information from various sources... whom should one hold responsible for a robot killing someone? The person who constructed the robot, the one who made the knife or the one who set up the computer programme?[6]

Flusser, like Metzger, a refugee from the Holocaust who lost his family in Nazi Concentration Camps, spells out these profound implications:

> A situation in which designers do not address themselves to these questions can lead to a total lack of responsibility. This is not a new problem, of course. It became terrifyingly apparent in 1945 when it was a question of deciding who was to be held responsible for the crimes against humanity committed by the Nazis. At the time of the Nuremberg trials, a letter written by a German industrialist to a Nazi official was discovered. In it, the industrialist timidly begs to be forgiven for having constructed his gas ovens badly: instead of killing thousands of people at one go, only hundreds were being killed… we find ourselves in a situation of total irresponsibility towards acts resulting from industrial production.[7]

Metzger has similarly spoken out against displacement and the industrial machine:

> Never before has the body been so threatened as now. The body is taken on by an emergent ever-enlarging assembly of instrumentation, surrounded by surrogates who replicate and replace the body. The human being is locked into a technoid double and becomes a mere template of that machinal self.[8]

Whilst the conceptual shifts made since the preparation of *Thought Pavilion* have been substantial, so too has been the process of identifying and aligning the core materials for this new work. All involved have engaged in some hard labour towards the production of an at times uncertain work. This has been very much its compulsion. New ground has been forged. Numerous art projects have taken place in the context of and with the field of robotics but few if any have combined perception science, robotics and stone. It has been an extraordinary undertaking. Latent moral questions not least about design and automation have added also to this potent mix.

Gilchrist and Joelson have spent months researching and testing the relative hardness and affordability of various types of stone. For months they closely considered Italian marble until finally realising that the more native Portland Stone was ideal for this particular task. Locating the right robot, negotiating its use and developing software equal to this aesthetically and philosophically challenging task has also been time-consuming.

Time has been a crucial factor within many aspects of the production not least the sense of uncertainty in relation to how long it might take the robot to actually achieve the sculpting of the void and also, in a previously uncharted experiment, to work with Metzger to achieve an optimal state of focus in thinking of nothing. According to Gilchrist and Joelson this took up to 20 minutes per session—a significant point if so as it has been pointed out by Kristin Stiles that the number 20 has been an important one in Metzger's life speculating that destruction of his first canvas using acid took 20 seconds and that:

> Metzger formulated his theory (Auto-destructive art) precisely 20 years after he was sent to England as a child of 12 in 1939, following his family's arrest by the Gestapo in Nuremberg. 20 seconds then is a temporal analogue for the time it took to destroy his personal world by killing his family; 20 years, the time of gestation in his own auto-transformation.[9]

With their decision to design a quasi auto-productive role for the chosen robot, as well as invite Metzger to be part of the *dramatis personae*, London Fieldworks have also opened up further significant lines of ethical enquiry as well as opening up potential for us to get a further set of insights into the mind of this important artist—not least the interplay between formal (including sculptural) elements in his work and his dialectical engagement with socio-political and 'deep ecological' concerns.

As an acclaimed progenitor of notions of auto-destruction (and auto-construction), enacted and depicted indelibly in his work since the late 1950s, Metzger has also throughout his career been a profound advocate of the dangers of the denial or neglect of a moral human compass with respect to our engagement with each other, the natural world and potential technological 'progress'. In a recent conversation with Jo Joelson and myself he makes the connection to what might be viewed as the force behind his conviction of personal responsibility or duty: i.e. his deep belief that as a Jew, with a direct responsibility to God, if he makes any wrong move there will be terrible consequences for the world.

Metzger, Gilchrist and Joelson live close to the former site of Loddiges which was renowned in the first half of the nineteenth century as the largest hothouse in the world, providing seeds and plants to the Crystal Palace, Kew Gardens etc.. Nearby Mare Street was the location of an arboretum with reputedly the largest managed tree collection in the world. It is a fit context for these artists who share a passionate commitment to what has been termed 'deep ecology', where the human's right to manage the world in his own image is held open to question. Their approaches to these issues are complementary.

London Fieldworks' signature works often attend to place and to habitat, building and sustaining interdisciplinary architectures and structures of engagement; investigating the meeting points of culture and nature through constructed environments and multi-sensory installations. A recent exhibition of film/sound work by the duo shown at the Ruskin Gallery in Cambridge, reflected on the work and beliefs of the rain-prophets of Northeast Brazil with respect to ecological balance.

Metzger's writings touch also on a note of prophecy; many recently warn of impending climate crisis, particularly the irreversible loss of species to man-made forces of extinction. Metzger has not fought shy with this in taking extreme positions, as embodied in his question published elsewhere in this catalogue when he asks if "the best way out for nature as a whole is the self-abolition of humanity?"[10]

The power of such statements is matched only by the creation of visually compelling, starkly stunning works such as *Flailing Trees* commissioned in Manchester in 2009 and shown in his Serpentine Gallery Decades show, also in 2009, which demonstrated taking principles of auto-destruction into phenomenology of change in the natural world. Asked why his works sometimes create pollution to draw attention to pollution Metzger has said:

> I compare this to homeopathy, which puts poison into the system in order to generate energy that could defeat the weakness, or the illness—in this case, the homeopathic dose of pollution. But doing all of this is not just a demonstration: it is a form; it's a creation, an attempt to show a visual experience that is unavailable except through destruction or pollution. And it's a very beautiful experience.[11]

In 1969, writing in *Studio International* in a fascinating article about Automata, Metzger observed:

> Great art may result from a total integration with mass-destructive drives in our societies... the "technology of paradise" is a step towards the salvation of the world... the conflict of artist and machine is entering a very critical phase... for the artist, formed on the Ruskin/Morris axis, our technology is a defeat....Very often a defeated subject comes as close as possible to that force that has defeated him.[12]

James Leach, a social anthropologist and among the leading critical commentators on contemporary art, science and technology relations wrote in 2005:

> Culture and technology have become so interdependent in current perceptions that the human project of civilisation and social development appear, to some, in danger of being hijacked by what Langdon Winner has dubbed "autonomous technology".[13] [14]

He cites Edmund Leach, an anthropologist writing in 1968, who said "the very means that humans have developed for controlling and utilising the natural world seem to have taken on a momentum of their own".[15]

In the *Studio International* essay mentioned above as well as describing computers as "becoming the most totalitarian tools ever used on society" Metzger reiterates the challenge thrown out by László Moholy-Nagy decades before: "This is our century: machine-technology-socialism. Come to terms with it, and shoulder the tasks of our century."[16]

His modernist commitment to principles of negation becomes here, with *Null Object*, fully realised—as the work attempts a formal realisation of the absence of self, surrendering to the presence of absence, becoming an outline of nothing. Whilst this issue of agency is apparent along the whole value chain of the project, it is for me most focussed around the figure of Metzger. The work also turns for me on the processes of auto-destruction being reviewed in the context of a process of collaboration between man and machine, through the ambiguous medium of software.

Having signed up to the displacement of his mind as the initiating human subject—in so doing performing a double absence or a double negation (thinking of nothing and removing his 'hand' or agency from the material production process)—his agency (in absence) then presents us with the vital questions about the focus or object of the work. Has he in a cerebral sense produced 'source code', and in keeping with free and open software practices, let it go? Metzger has told me that in terms of control and agency in his Auto-destructive art works it was a very important thing to know how 'much to let go'.[17] Here, he takes part as a subject or restrained agent in a work which is not his own, with his 'thought of nothing' becoming the source or control measure against which all the other inputs will be gauged. As always with his work, this measuring out of a capacity to relinquish control leads onward to questions of ethics and responsibility. In this willingness to be 'subject' here, Metzger's childhood inevitably is evoked.

Born an Orthodox Jew in 1926, with wide open eyes and ears he was a young person witnessing at first hand in Nuremberg the march of Nazi boots in the streets, the rallies and the cinematic events which demonstrated the power of media to amplify, to sensationalise and to control minds by exercise of spectacle, focus and symbol.

Speaking at a roundtable event in Glasgow on a theme of 'self-cancelling' held in 2008, Metzger said:

> Talking of migrating butterflies losing momentum and losing substance, it occurs to me that as a child I would watch tens of thousands, hundreds of thousands of Nazis in variegated uniforms, either the SA, the SS, or the Army, marching every year in Nuremberg, the town where I was born and where I lived until the age of 12, nearly the age of 13 at the beginning of 1939 when I left and came to this country. And so year after year there would be the annual Nuremberg Rallies, the most important meetings of National Socialism, year by year from 1933 I would watch these armies, these endless columns of marching people for one or two weeks and we lived just off the main road, the biggest road in Nuremberg, leading from Nuremberg to Fürth which is miles and miles and miles, marching, marching.[18]

Metzger and his brother Mendel had been sent to England in 1939. Metzger's parents sadly perished in 1943. His elder brother also died at the hands of the Nazis.

Having learned to be a furniture maker in Leeds in the early 1940s, Metzger moved to Bristol for a brief time in 1944 to join an Anarchist/Trotskyist collective. He recounts how Trotsky's death in 1940 made a considerable impact on him, as did that of Eric Gill in the same year. Whilst living in the collective, he made his decision to go the direction of art, initially considering sculpture, then on the advice of Henry Moore studying life-drawing which he did in Cambridge, London and Antwerp from 1945 into the 1950s. He left England temporarily in 1948 after a period which he recalls as a particularly difficult stage of his life; travelling with his brother in Belgium and France he found himself able to find a way forward intellectually and personally. On return to England Gustav had his grant renewed to continue his evening class studies

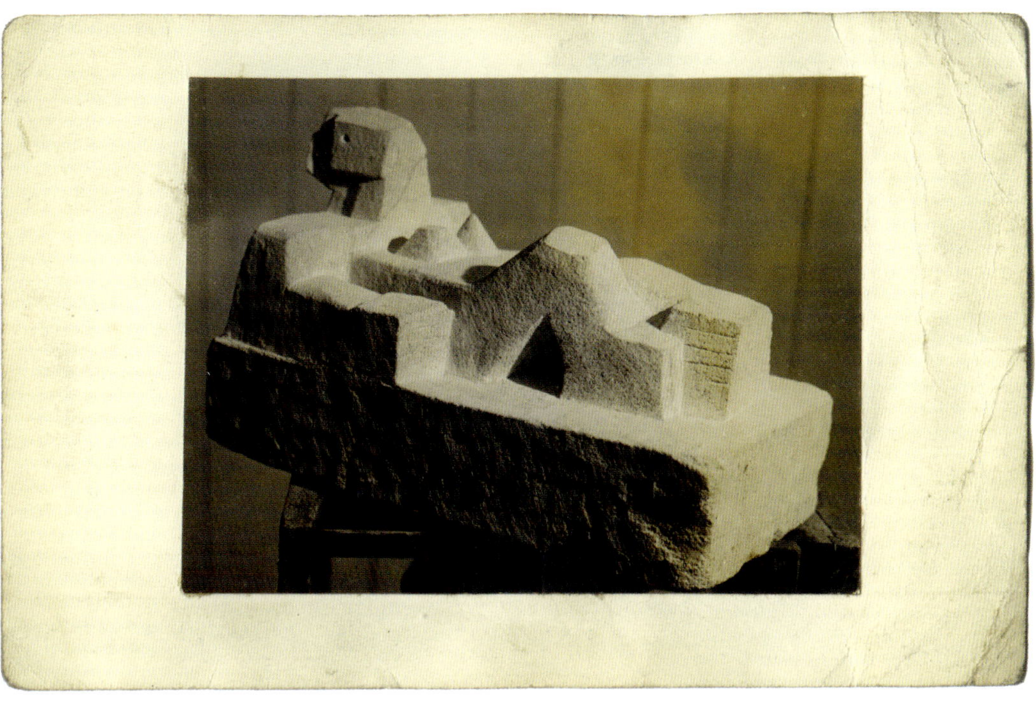

Untitled
Early work by Gustav Metzger
Courtesy the artist

at the Borough Polytechnic (until the summer of 1953) with David Bomberg, another great Jewish educator and artist. Sculpture was of major importance to him. He has spoken of being inspired not least by Henry Moore, Eric Gill and Jacob Epstein who in his view was not only a great artist but also a great religious one. That Gill and Epstein who were both close associates is fascinating to note here not least because of their direct approach to materials and to experimentation with techniques often called 'primitive':

> In Jacob and the Angel, Epstein 'seems to have tapped the mysterious source of energy that so often animates primitive sculpture, without imitating any actual features' (catalogue Tate Gallery exhibition 1961). The use of this primitivist style when dealing with religious subject matter was found shocking by many of Epstein's contemporaries.[19]

Gill also saw himself as a primitive, for years carving directly onto the stone. His *Stations of the Cross* for Westminster Cathedral were both loved and strongly criticised in equal measure for their plainness.

The controversy associated with primitivism may be difficult to appreciate now. A primary goal of the *Null Object* research has been to try to locate it and to explore the brain's interaction with it.

At lunch before sundown this year on Yom Kippur Metzger spoke movingly to Jo Joelson and myself of the happiness of his childhood, the herrings and chestnuts his family would eat in winter in the beautiful old streets of Nuremberg and of how he would sit in the cinema for hours watching contemporary films including those being made in and about Germany at that time such as the work of Leni Riefenstahl (and including of course *Olympia*, her film of the 1936 Berlin Olympics). He also reminded us of the sense of duty and responsibility instilled in him in a religious Jewish household. Asked what Yom Kippur now meant to him, he spoke of how it was a day to let business go and to renew the roots of one's relationship with Judaism. The painting *The Scapegoat* by Henry Holman Hunt for him represented this perfectly. He explained how he had been taught that "If I make one mistake the entire world might collapse; we (Jews) must not underestimate the danger of what we are doing or about to do... we have to change to make sure we do not destroy the world by a single wrong act— which explains my concern with extinction."[20]

Living in England, mostly in King's Lynn in Cambridgeshire, during the 1950s, he was grappling for years with the central question: how might a Jewish person make art? He has spoken of how he was circling Judaism with respect to its prohibition on making images of the human figure. He made a particular work repeatedly, which was a painting of a table. He has recently described being preoccupied for a long time with this work, doing many sketches and from the elongations of the process of working on the images of the table eventually what emerged was a fairly large painting which also held the image of an enormous atomic mushroom cloud. This was bought by a member of the Committee of 100 who left London to settle in the Caribbean in the 1960s where hopefully one day it will be found. About his series of paintings he suggests "a connection to the tablet of Moses...".[21]

With this we might see the complexity of the problems with which Metzger was grappling for resolution through art. The quality and potency of Metzger's drawings from the period of the 1950s have recently been recognised with a major display at the 2012 Kassel Documenta. Whilst he is well-known as a great artist from the 1960s it is only now that the development and fermentation of his work and ideas in the preceding decade is receiving due attention partly because a collection of his work from that time was only recently uncovered from storage. Their preservation had been in question so this is a major recovery of

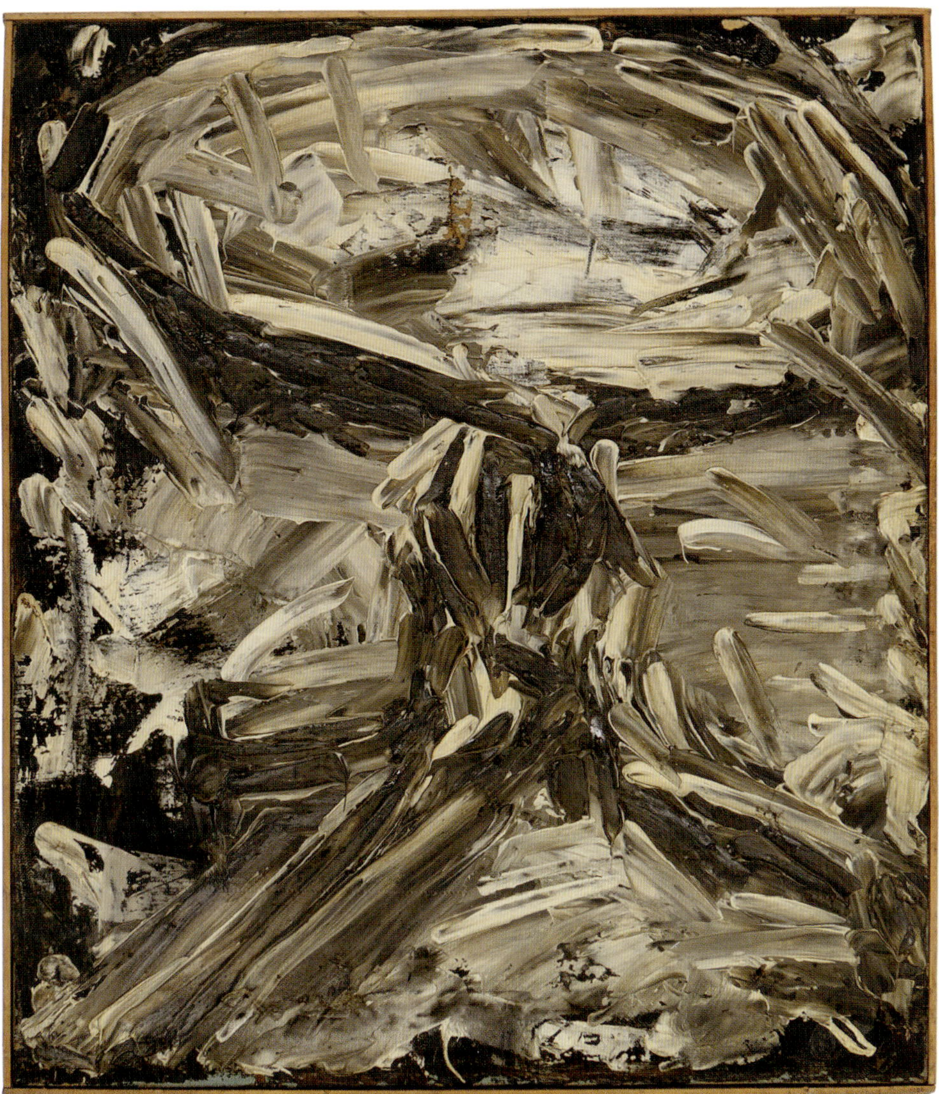

Study for *The Table*
Gustav Metzger c. 1957
Courtesy the artist

significant works that are antecedent to those for which Metzger is better known.

Metzger had joined CND in the Eastern region in 1957 and became a prominent figure in the anti-nuclear movement, joining in the Aldermaston March of 1959 and the Committee of 100 which Metzger himself helped name and which included Bertrand Russell and many other leading British intellectuals. The Committee of 100 helped to build a campaign of civil disobedience adopting tactics of resistance through peaceful public protest, influential to many later campaigners and activists.

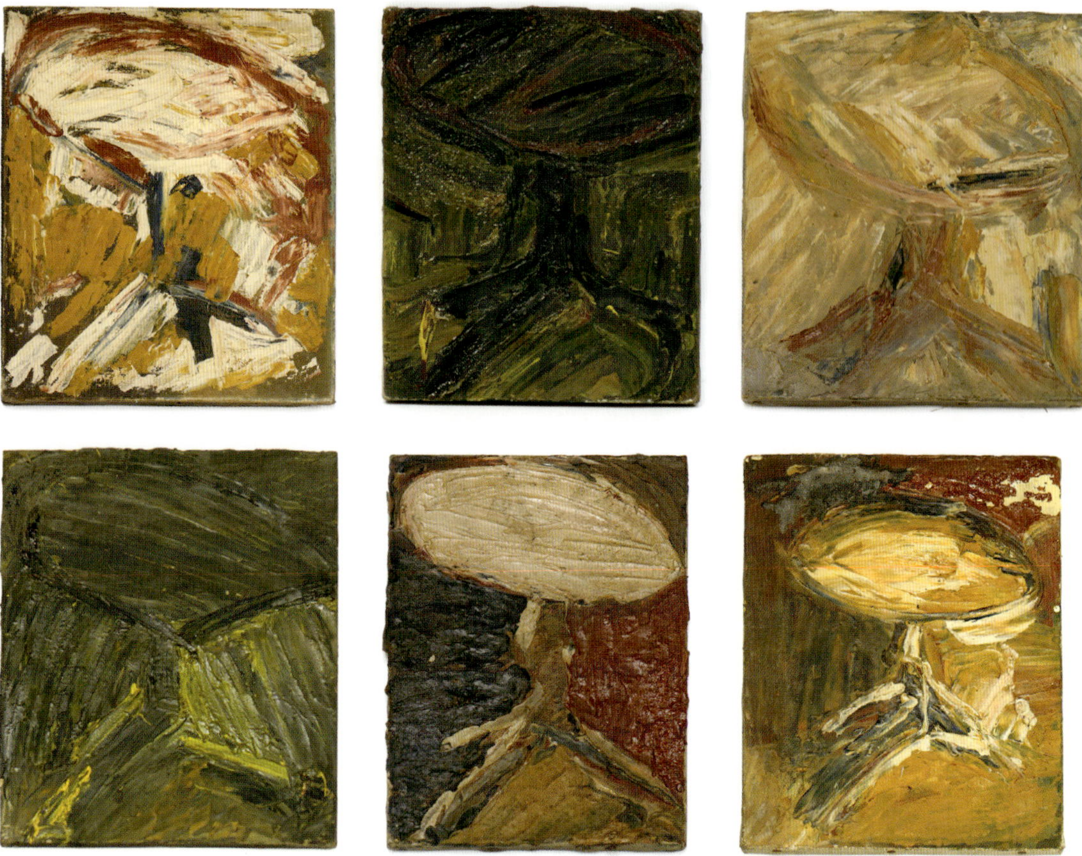

Studies for *The Table*
Gustav Metzger c. 1957
Courtesy the artist

In his philosophical writing, Bertrand Russell engages also in discourse around the difference between materiality and the mind which reveals many preoccupations relevant to *Null Object*:

'Whatever can be thought of is an idea in the mind thinking of it, therefore nothing can be thought of except ideas in minds;' … Now obviously this point in which the philosophers are agreed—the view that there is a real table, whatever its nature may be is vitally important, and it will be worth while to consider what reasons there are for accepting this view before we go on to the further question as to the nature of the real table…. It has appeared that, if we take any common object of the sort that is supposed to be known by the senses, what the senses immediately tell us is not the truth about the object as it is apart from us, but only the truth about certain sense-data which, so far as we can see, depend upon the relations between us and the object.

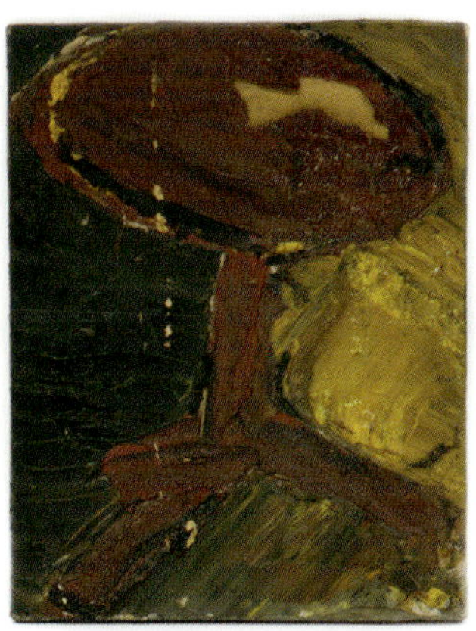

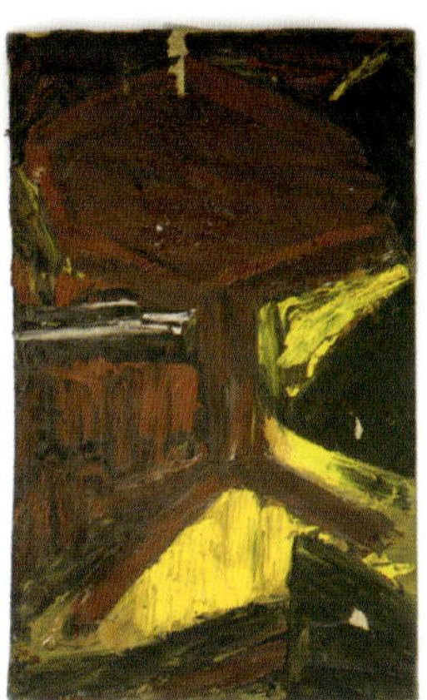

... Thus our familiar table, which has roused but the slightest thoughts in us hitherto, has become a problem full of surprising possibilities. The one thing we know about it is that it is not what it seems. Beyond this modest result, so far, we have the most complete liberty of conjecture. Leibniz tells us it is a community of souls: Berkeley tells us it is an idea in the mind of God; sober science, scarcely less wonderful, tells us it is a vast collection of electric charges in violent motion. Among these surprising possibilities, doubt suggests that perhaps there is no table at all.[22]

Having spent many years confronting the challenges of making art, within the constraints as outlined above, Metzger realised by 1959 that he was "on my way to Auto-destructive art... Auto-destructive art came as a great release as I could make art without dealing with the human figure".[23]

In 1959 Metzger published his "Auto-destructive art" Manifesto and in 1961 made the Auto-destructive work depicted in an iconic photograph applying hydrochloric acid onto three stretched nylon sheets on the South Bank in London.[24] In its performance of destruction as well as handling of dangerous materials wearing a gas mask, there are clear references back to the Nazi gas chambers. The Sixth International Union of Architects Congress also took place on the South Bank in 1961 on the themes of New Techniques and New Materials. He made the work independently from this and it stands as critical to the frame of cultural references which may be applied to understanding the trajectory between post-war and 1960s Britain. His action sculpts meaning through its power of negative dialectics. It is a fitting memorial to the psychic wounds of the period: rendering an action in time and through time, a performative action that simultaneously places and replaces art and artist, exercising control of new materials—or what we might call new media—to a point where it was necessary to let go. An uncanny resemblance between the void in the stone and the gap in the canvas has also been observed.

The fascination of Metzger's work lies somewhere in this tension between control and release of control. As Paul Broks notes above, there are limitations to scientific processes of visualisation and analysis in informing us about the core motivations or desires which we might hold closest to ourselves. As Metzger described aspects of his childhood and how Judaism has been the epicentre of his life since boyhood, it is inevitable that one might consider perhaps at a deeper level the historical event which becomes 'cut out' or sculpted

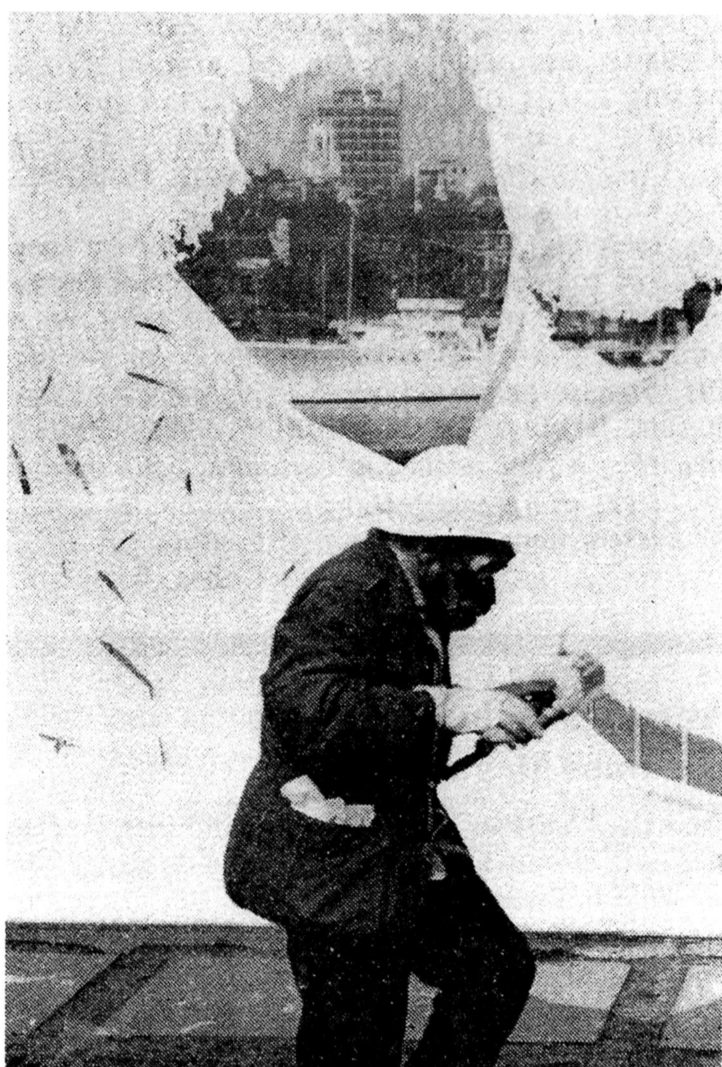

Photograph taken by
Mendel Metzger during the
filming of Gustav Metzger
giving a demonstration of
Auto-destructive art
1963
Courtesy Gustav Metzger

by the exercise of robotic power whetted by a constant onslaught
of water on stone in an Oxfordshire workshop. The presence of the
86-year-old artist in the *Null Object* work—like some subliminal
alchemical agent who also cuts away to the bare truth, the primitive
state, of the matter—achieves an affordance like an after effect, much
like the transfixing nature of the work on the South Bank decades
before. "Extinction is irrevocable. When it happens—it is over. It sums
up everything I am about to say."[25] Metzger has often said his concern
has been for the future not the past. Both seem fused in his work and
his life.

In *Artforum* in 1969, Jack Burnham wrote about his perception that:

Artists are "deviation amplifying" systems, or individuals who, because of psychological makeup, are compelled to reveal psychic truths at the expense of the existing societal homeostasis. With increasing aggressiveness, one of the artist's functions is to specify how technology uses us.[26]

Metzger's willingness to stand up for his beliefs was apparent when he went to jail for a month in 1961 after 35 members of the Committee of 100 were arrested before a CND demonstration for civil disobedience. In court he stated:

I came to this country from Germany when 12 years old, my parents being Polish Jews, and I am grateful to the government for bringing me over. My parents disappeared in 1943 and I would have shared their fate. But the situation is now far more barbaric than Buchenwald, for there can be absolute obliteration at any moment. I have no other choice than to assert my right to live, and we have chosen, in this committee, a method of fighting which is the opposite of war—the principle of total non-violence.[27]

Like the then 89-year-old Bertrand Russell, he chose jail rather than the offered terms of being bound over on good behaviour for a year.

Adorno in his *Negative Dialectics*, tells us that if thinking is to be true today it must be a thinking against itself, a thought that appears deeply relevant to Metzger's life and work, as has been made articulately by Andrew Wilson and others.[28]

Gilles Deleuze has described the force of auto-destruction in his readings of Friedrich Nietzsche's (esoteric) theory of the circle of eternal recurrence: "Auto-destruction is said to be an active operation, an 'active destruction'. It and it alone expresses the becoming active of forces, and forces become active to the extent that the reactive forces deny and suppress themselves in the name of a principle that, scarcely a moment ago, assured their conservation and triumph."[29]; in *Thus Spoke Zarathustra*, his great hymn to transformation Nietzsche wrote: "Hear it ye creating ones! Change of values—that is change of the creating ones. Always doth he destroy who hath to be a creator."[30]

Whilst orthodox Judaism tends towards belief in linear time, the idea of a circle of eternal return has its roots in so-called primitive cultures, with those who live arguably outside historical time with faith in the infinite recurrent cycles of nature or cyclical periodicity as Mircea Eliade has argued: "the myth of eternal return restores historical time… to a place in the time that is cosmic, cyclical and infinite".[31]

Metzger has also spoken of his interest in cosmology and the beginning and end of space, life and time: "This way of thinking is very close to Auto-destructive art. Now we accept the world started with nothing and that it will end with nothing but it is going through enormous transformations that involve destructivity."[32]

Null Object exposes lineaments of Metzger's and London Fieldworks' earlier work and current preoccupations. It does this by abnegation and by an implicit revealing. The emptiness and absence in the heart of the stone is preserved and preserving.

NOTES

[1] Old Testament, Daniel 2:34-35 (34). New International Version, 1984.

[2] Woolf, Virginia, *Between the Acts*, text first published by Hogarth Press, 1941 (excerpt at Google books accessed October 2012), p. 123.

[3] Bruce Gilchrist, in conversation with author during Digital Dancing, Riverside Studios, 1997.

[4] Palmer, Judith, "Welcome To The Pleasure Zone", *The Independent*, May 1998, www.londonfieldworks.com/projects/gastarbyter/press.php.

[5] Broks, Paul, in *Matter into Imagination*, a catalogue to the exhibition by Susan Aldworth at Menier Chocolate Factory, London, 2006.

[6] Flusser, Vilém, "Ethik im Industriedesign", copyright Bollman Verlag GmbH 1993, English language translation in *The Shape of Things, A Philosophy of Design*, London: Reaktion Books Ltd, 1999, p. 67.

[7] Flusser, Vilém, "Ethik im Industriedesign", *The Shape of Things, a Philosophy of Design*, p.68.

[8] From Jones, Alison, "Metzger's Historic Photographs", Forum for Holocaust Studies, www.ucl.ac.uk/forum-for-holocaust-studies/metzger.html

[9] From Stiles, Kristine, "Uncorrupted Joy: International Art Actions", in *Out of Actions: Between Performance and the Object 1949–1979*, P Schimmel ed., London: Thames & Hudson, 1998, p. 272. Cited at www.ucl.ac.uk/forum-for-holocaust-studies/metzger.html

[10] Metzger, Gustav, "Empty and Emptiness, and Emptying", in *Heft Der Leeren*, M Copeland ed. for the Kunsthalle Bern to accompany Voids, Eine Retrospektive, 2009. Also see Gilchrist, Bruce, Jo Joelson, *Null Object: Gustav Metzger thinks about nothing*, London: Black Dog Publishing, pp. 90–91.

[11] Cited in interview with Gustav Metzger by Mark Godfrey, first published in *Frieze* magazine, Issue 108, Jun–August 2007. www.frieze.com/issue/article/protest_and_survive/

[12] Metzger, Gustav, "Automata in History", *Studio International* journal, March 1969, pp. 107–109. www.bbk.ac.uk/hosted/cache/archive/CAS/Metzger,%20G.%20Automata%20in%20history.%20Studio%20International.%20March%201969%20p107-109.pdf

[13] Leach, J, "'Being In Between' Art-Science Collaborations and a Technological Culture", *Social Analysis*, vol. 49, no. 1, Spring 2005, pp. 141–160.

[14] Winner, Langdon, *Autonomous Technology*, Cambridge, MA: MIT Press, 1977.

[15] Leach, Edmund, *A Runaway World? The Reith Lectures 1967*, London: BBC, 1968.

[16] Moholy-Nagy, László, in *Experiment in Totality* by Moholy-Nagy, Sybil, New York: Harper & Brothers, 1950, p. 18. Full text accessible at www.archive.org/stream/moholynagyexperi007463mbp/moholynagyexperi007463mbp_djvu.txt

[17] Citing Metzger, in interview with author, 11 September 2012.

[18] Gustav Metzger & Self-Cancellation: Round Table Discussion, Chair Brian Morton, Mackintosh Room, Glasgow School of Art, Saturday 16 February 2008. www.artandresearch.org.uk/v3n1/metzger.html

[19] Rothenstein, John in Epstein, exhibition catalogue, London: Tate Gallery, 1961. www.tate.org.uk/art/artworks/epstein-jacob-and-the-angel-t07139/text-summary

[20] Conversation with Gustav Metzger, Jo Joelson and author, 26 September 2012.

[21] Conversation with Gustav Metzger, Jo Joelson and author, 26 September 2012.

[22] Russell, Bertrand, *The Problems of Philosophy*, Home University Library, 1912.

[23] Conversation with Gustav Metzger, Jo Joelson and author, 26 September 2012.

[24] Metzger, Gustav, "Auto-Destructive Art", 1959, in *Theories and Documents of Contemporary Art: a sourcebook of artists' writings*, Kristine Stiles and Selz, Peter eds, Berkley, LA and London: University of California Press, 1996, p. 402.

[25] Conversation with Gustav Metzger, Jo Joelson and author, 26 September 2012.

[26] Burnham, Jack, "Real Time Systems", *Artforum*, 8:1, September 1969, pp. 49–55. www.wallflowermedia.com/art2704/RealTime-BURNHAM.pdf

[27] Gustav Metzger quoted in Wilson, Andrew, "Papa what did you do when the Nazis built the concentration camps? My dear they never told us anything", Metzger, Gustav, *Damaged Nature, Auto-Destructive Art*, London: Coracle Press, 1996, p. 73. Quoted in Jones, A, "Introduction to the Historic Photographs of Gustav Metzger". www.ucl.ac.uk/forum-for-holocaust-studies/metzger.html

[28] Wilson, Andrew, "Gustav Metzger: A Thinking against Thinking", *Gustav Metzger: Retrospectives*, Museum of Modern Art Papers, vol. 3, I Cole ed., Oxford: MOMA, 1999, p. 72.

[29] Deleuze, G, "Active and Reactive", in *New Nietzsche: Contemporary Styles of Interpretation*, David B Allison ed., Cambridge, MA: MIT Press, 1985, pp. 101–192.

[30] Nietzsche, Friedrich, *Thus Spake Zarathustra—A Book for All and None*, Thomas Common trans., Edinburgh: TN Foulis, 1909. www.gutenberg.org/files/1998/1998-h/1998-h.htm

[31] Eliade, Mircea, *The Myth of the Eternal Return: or Cosmos and History*, New York: Bollingen Foundation Inc., 1954.

[32] Ulrich Obrist, Hans, *The Conversation Series, Number 16—Gustav Metzger*, Köln: Verlag der Buchhandlung Walther Konig, 2009, p. 34.

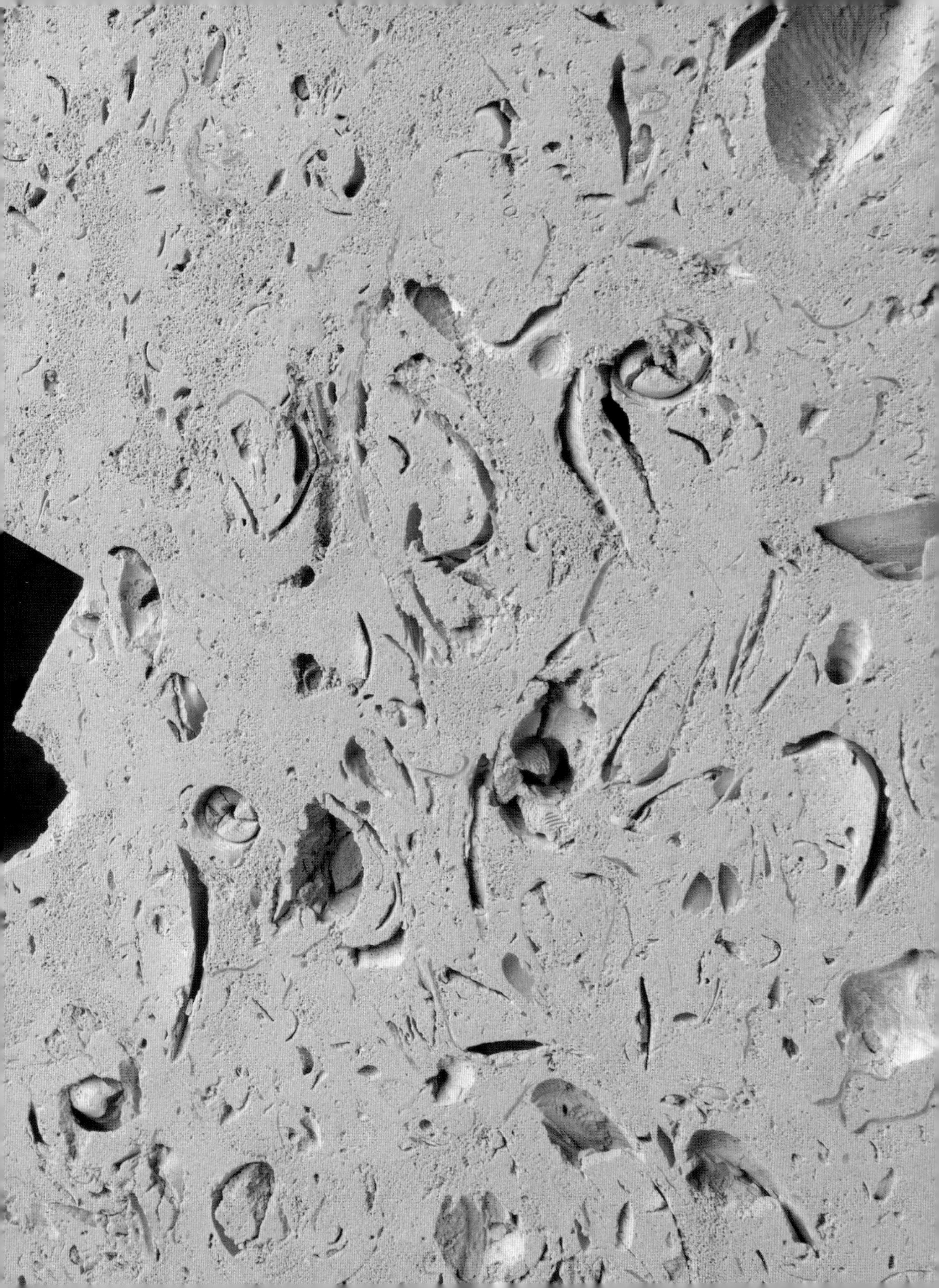

WRITING ABOUT NOTHING

Hari Kunzru

LEAR: ... what can you say to draw a third more opulent
than your sisters? Speak.
CORDELIA: Nothing, my lord.
LEAR: Nothing?
CORDELIA: Nothing.
LEAR: Nothing will come of nothing, speak again.

[*King Lear* I.i]

Lear speaks out of an ancient tradition of Western thought. Nothing will come of nothing: Creation cannot proceed *ex nihilo*. The Greek model of creation proposes a material substrate, known as chaos, which is operated on by the gods. Later monotheistic theology identifies the agent forming the formless stuff of chaos as a singular God. Later still, for Christianity, Islam and Judaism, God is held to create out of himself, without working on some pre-existing material. As God is total and infinite, creation *ex Deo* doesn't arise out of nothing; It arises out of everything. Throughout the history of Western thought, the idea of nothing is always bound up in this negative way with that of creation. The opposition of the two has the force of a self-evident truth: you can't make something out of nothing. You just can't.

For Lucretius, "In de Rerum Natura" (The Nature of Things) "the first of Nature's basic principles" is that "no thing can ever be produced by the gods from nothing". Instead of creation *ex nihilo*, there are 'seeds' which, unfolding in time and space, give rise to everything that exists. He warns that if creation could proceed out of nothing:

any kind of creature
could then be born from anything, with no need for seeds.
Men would come from the sea, and birds and scaly fishes
arise from the earth.

Creation from Nothing would be lawless, senseless. It would take place outside the temporal laws of the normal seasonal cycle, outside the spatial laws of the growth of the particular organism from its seed. Lucretius puts the idea of nothing to work in the project of defining the existent world as a fundamentally lawful place. This is a very successful, productive hypothesis. It marks, among other things, the origin of modern scientific enquiry.

However, at the very origin of the Western philosophical tradition there is a tear in the membrane between something and nothing, between lawful existence and lawless absence. In one of the few surviving fragments of his poem "On Nature", Parmenides writes: "It must be that what can be spoken and thought is; for it is possible for it to be and it is not possible for what is nothing to be." If "what can be spoken and thought is", then mental phenomena must be real. But what kind of reality do they have? It's a question which haunts Western thought to this day. What ontological status to give to dreams, trance and in particular to the imagination, which appears to have the power to create *ex nihilo*?

The claim that dreams, creative states and meditative states are ineffable, intrinsically anti-rational, or just 'personal' and not be available for scrutiny, forms a kind of curtain that many people, particularly artists and religious believers, prefer to draw across their mental processes. The threats are manifold—the banalisation inherent in a scientific explanation of consciousness, the possibility that a process or experience that 'should be' unique will be replicated, loss of identity. The dominant cultural idea of artistic creativity (still) comes from Romanticism, with its flamboyant image of the artist heroically wrenching art out of his subjectivity, each one a little action-figure of God creating *ex Deo*. Contemporary neuroscience has moved towards a model of populations of mental processes, in which our experience of self, let alone heroic creativity, has been relegated to the status of a side effect, or at best some kind of component subroutine of behaviour.

For Lucretius, writing at the origin of scientific enquiry, all mental phenomena are unequivocally something rather than nothing. They are material, lawful and hence potentially knowable. He decides that the "images" which "terrify our minds, either when we are awake or when we are sleeping, when we catch a glimpse of strange shapes and phantoms of the dead and departed" are "a sort of outer layer peeled off an object's surface". He describes them in terms of dust, smoke, the skin shed by a snake. Unfortunately mental "images" have a lot of the absurd characteristics of creation out of Nothing. There is no apparent temporal or spatial regulation in the mind, which seems to possess a kind of seedless, chaotic fecundity, and can easily produce monsters, like "livestock bursting forth from the sky" and "men of such a size that they could wade across the deepest sea, as a ford". Mental processes are infected by hollowness, by non-existence. They seem to bear a disturbing relationship to impossibility, and hence to nothing.

For many years, Bruce Gilchrist has attempted mindful crossings of the boundary between something and nothing. His persistent interest in shamanism, meditation, dreaming and artistic creativity is infused with a sort of Lucretian materialism. In his solo work and his more recent collaborations with Jo Joelson, impressions are often transmitted directly, sensations thrown off bodies like diffuse shells or 'skins'. He and Joelson have experimented with EEG, skin galvanic response and other 'direct' ways of recording physiological and neurological states. The work almost always involves the digital storage and reproduction of those states. Can one person's experience of meditation be reproduced in another? Could it be visualised? How about the pain of having a tattoo? Where other artists see banality or threat, Gilchrist and Joelson see a sort of porousness.

To computer programmers, a null object, invoked in the title of Gilchrist and Joelson's latest project, is useful because it does nothing. It has no behaviour. It represents nothing, but it is not nothing. It is a nothing which is something, which can be referenced, made useful. The project of asking an artist to think about nothing, and using this as the starting point of an act of creation (and not just any creation, but heroic sculpture in Portland Stone) is gorgeous in its perversity and precise in the questions it poses. Choosing Gustav Metzger as the artist who will be doing the thinking opens *Null Object* out into a radical tradition of post-war aesthetics that challenges the legitimacy of a contemporary art establishment obsessed ever more crassly with product and celebrity, with a definition of value as presence and permanence. Unless it will stoop to becoming a commodity, nothing has no place in a gallery.

Metzger is best known as the initiator of DIAS, the 1966 Destruction in Art Symposium, which involved artists such as John Latham, Yoko Ono and Barry Flanagan and introduced London to the work of Viennese Actionists Herman Nitsch and Otto Muehl. DIAS presented a wide variety of forms of art that enacted or in some other way contained its own annihilation. In actions during the late 1950s and early 1960s he wore protective clothing and sprayed acid onto nylon sheets, which gradually dissolved. Seen in a 1963 film, (*Auto-Destructive Art—The Activities of G Metzger*, by Harold Liversidge), the busy gestures of this small intense man, dressed as if for chemical or nuclear attack, seem to undercut the macho heroics of Abstract Expressionism. Metzger has frequently expressed his wish to decentre the figure of the artist as original creator. Interviewed for a documentary about Michael Landy, he faced away from the camera. "I am", he said, "opposed to the artist as celebrity."

Destruction is not Metzger's only gesture of disgust at the narcissism of contemporary art stardom. He has also followed interests in what might be termed bottom-up or emergent creation such as crystal growth. For long periods, he has refused or tried to refrain from making art at all. Hooked up to Gilchrist and Joelson's network, he will continue this activity. He will try to think about nothing, to make nothing. The art will be a byproduct of that activity; it will come into being despite the effort of the artist, not because of it. To be precise, the negative space carved into the block of stone will be created by Metzger's involuntary imagination of primitives, shapes specified by Gilchrist and Joelson's software. These shapes will provide the control instructions for the CNC (computer numeric control) robot, which will carve the block. Heroic sculpture, the most macho medium of them all, 'achieved' by failure.

To begin to answer the question posed by *Null Object* (to see if it is even a meaningful question to ask) we need to have a better sense of what thinking about nothing really entails. To think about nothing is, of course, to think about some idea of nothing, a Null Object, a domesticated pseudo-nothing which can only ever be a stand-in for the absolute. Parmenides: "what can be spoken and thought is". Absolute nothing must, according to this ancient logic, be defined as that which is essentially unthinkable, unknowable—or if even this faint whiff of reification is too much, then as a kind of horizon to thought, the point at which all thought's sailing ships fall off the edge. In many mystical traditions this cosmic absence is itself identified with God, and the purpose of meditation is a negation of thought, an unravelling of the thinking, intentional self. Our idea of Nothing, whatever it might be, is always a sort of screen over this radically-unknowable, unreferenceable non-object. We think of a nothing which is secretly something, which we covertly put to work, doing whatever conceptual business we require.

So can I write anything at all about nothing? Or is this essay going to end up as a longwinded retelling of the story (quoted here in full) of the red-haired man, written in 1937 by the Russian absurdist Daniil Kharms?:

> There was a red haired man who had no eyes or ears. Neither did he have any hair, so he was called red-haired theoretically.
> He couldn't speak, since he didn't have a mouth. Neither did he have a nose. He didn't even have any arms or legs. He had no stomach and he had no back and he had no spine and he had no innards whatsoever. He had nothing at all! Therefore there's no knowing whom we are even talking about.
> In fact it's better that we don't say any more about him.

In the nonsensical tale of the subject which has no attributes (and hence vanishes, like the victim of a Stalinist purge) is a bitterly ironic reference to the mystical tradition of the *via negativa* (the negative way), which proposes subtraction as a sort of mental technique, a road towards discovery of God as a pure entity, without appearance or form.

On the same maner schalt thou do with this lityl worde God. Fille thi spirit with the goostly bemenyng of it withoutyn any specyal beholdyng to any of His werkes whether thei be good, betir, or alther best, bodily or goostly—or to any vertewe that may be wrought in mans soule by any grace, not lokyng after whether it be meeknes or charité, pacyence or abstynence, hope, feith, or sobirnes, chastité or wilful poverté. What thar reche in contemplatyves? For alle vertewes thei fynden and felyn in God; for in Hym is alle thing, bothe by cause and by beyng. For hem think and thei had God, thei had alle good; and therfore thei coveyte nothing with specyal beholdyng, bot only good God. Do thou on the same maner, as forth as thou maist by grace; and mene God al, and al God, so that nought worche in thi witte and in thi wile, bot only God.

The Cloud of Unknowning [ch. 40]

The Christian mystic of this anonymous fourteenth-century English text seeks, by contemplation of nothing, union with an ineffable God. In this project, his enemy is the imagination, which unless it:

be refreyned by the light of grace in the reson, elles it wil never sese, sleping or wakyng, for to portray dyverse unordeynd ymages of bodely creatures; or elles sum fantasye, the whiche is nought elles bot a bodely conseyte of a goostly thing, or elles a goostly conseyte of a bodely thing. And this is evermore feynid and fals, and anexte unto errour.

[ch. 65]

The student of the Tibetan *Bardo Thodol* (the Great Liberation on Hearing in the Intermediate State) seeks in the suppression of imagination and the contemplation of nothing an escape from the temporal cycle of birth, death and rebirth. The *Bardo Thodol* is essentially a technical manual, designed for a practitioner to chant to the dying person, to teach them how to recognise reality and so liberate themselves from falling back into rebirth. Each mental state (waking, meditation, dream, death, rebirth) is a bardo, an "intermediate state" between other states, each of which offers opportunities for learning and the danger of deception. After the Bardo of the Moment of Death arises the bardo of reality. The practitioner speaks to the dying person:

O Child of Buddha Nature (call the name of the dying person) the time has now come for you to seek a path. As soon as your respiration ceases, [the luminosity] known as the 'inner radiance of the first intermediate state', which your spiritual teacher formerly introduced to you, will arise. [Immediately] your respiration ceases, all phenomena will become empty and utterly naked like space. [At the same time] a naked awareness will arise, not extraneous [to yourself], but radiant, empty and without a horizon or centre. At that moment, you should personally recognise this intrinsic nature and rest in the state of that [experience].

If the dying person fails to recognise the intrinsic nature of reality (and the illusory nature of self as separate from this reality), he or she will fall steadily backwards through a series of post-death bardo states, each of which is more cosmically terrifying and distracting than the last. By the eleventh day after death "he who is called the transcedent lord Padma Heruka, of the Padma family of blood-drinking deities, will arise from the Western direction of your brain...". Each time, the Child of Buddha Nature is offered the chance to recognise reality: "radiant, empty and without a horizon or centre", until he finally falls back into a body and is reborn.

The mental state required to recognise reality and stay in it is generally translated into English as "Buddha-mind". The qualities of this so-called "pristine cognition" are given in the writing of the Nyingma school as "manifest enlightenment", "indivisible indestructible reality", "great sameness", "great non-discursiveness" and "liberator of sentient beings".

In *Null Object*, a negative space is carved in the block of stone despite Metzger's attempts to achieve pristine cognition. It is specified by comparing his EEG readout to a library of EEGs collected from volunteers who have been asked to watch a screen on which a 3D primitive emerges out of a stereogram image. The moment when the shape is perceived by the subject is captured and added to the library. As Gilchrist describes it:

> we have been working with 20 minute EEG recordings of Gustav attempting to think of nothing. 20 minutes of data—sampling the relational database every two seconds—gives us 600 primitive objects, which is enough to make the void. Our software rotates and collides the primitives. If one is a close match to the database, it appears smaller and closer to the edge of the block of stone—giving the internal surface of the stone a finer grain because all the smallest objects have been moved there. If a primitive is the result of a distant match, it appears larger and closer to the centre of the block of stone.

If one reads the *Bardo Thodol* as a materialist text, describing in a Lucretian manner real, lawful, existing phenomena, it's hard not to feel that the two things—the complex, recursive shape carved into the block and the hosts of gods and demons which beset the mind in the bardos of death—are actually the same thing. The carving is a sort of print-out of demons, a trace of the terrible chaos from which the meditator seeks liberation through the contemplation of nothing.

But who thinks of nothing? Who is performing this task? This is perhaps the most challenging question raised by *Null Object*. So far, in accordance with the canons of Romantic aesthetics, Metzger's activity has been framed as an act of will, the heroic struggle of a conscious subject, albeit one that is seeking its own dissolution. In 1983 a physiologist called Benjamin Libet published a paper that remains central to neuroscientific thinking about free will. Libet asked his subjects to flick their wrists at some random moment, while he monitored the associated activity in their brain, specifically an electrical signal in the motor cortex known as "readiness potential", which leads up to voluntary muscle movement. He found that this activity began up to half a second before the subject had the conscious awareness of intending to move. Though the "Libet experiment" has been criticised in various ways, its main conclusion—that the electrical preparation for typing these words takes place before I am aware I've decided to move my fingers over the keyboard—still holds.

The authors of a 2008 paper that concludes, dramatically, that "the outcome of a decision can be encoded in the brain activity of the prefrontal and parietal cortex up to ten seconds before it enters awareness" hypothesise that "this delay presumably reflects the operation of a network of high-level control areas that begin to prepare an upcoming decision long before it enters awareness". These "high level control areas" are, according to the current neuroscientific model, themselves merely nodes on a distributed network, nodes that are interacting dynamically with each other to initiate and govern mental processes. This network is composed of populations of neurons. Populations, all the way down. No unitary willing self, acting decisively on the yielding raw material of the world. Sorry, Ayn Rand.

The thinking subject is a swarm, a horde. The idea of a sovereign self, sitting at the 'controls', has been consigned to history. The attempt (still common) to localise the moment when some mythical 'conscious' self enters the decision-making process, taking over the reins from some 'unconscious' or autonomic routine, also feels archaic, misguided. Conscious intention plays a role in decision-making, but its precise function is fiercely disputed. Daniel Wegner, a collaborator of Libet, prefers to think of the self, not as a boss, but a sort of "public relations agent", governing social interactions and providing a sense of continuity between past actions and anticipated intentions.

In *Null Object*, the responsibility for sculpting a block of stone is devolved from some mythical sovereign artist (Michelangelo, Rodin) to a network. Is it the artist Bruce Gilchrist? Jo Joelson? Is it the software which specifies the primitives, the volunteers whose EEGs form the library, a KUKA industrial robot, Harvard scientists, technicians at a fabrication plant in Oxfordshire, or Gustav Metzger? *Null Object* mocks the persistent narcissism of the artist, who believes secretly that he is a little god. It is a release into a more profound and complex reality. A great liberation.

REFERENCES

Anonymous, *The Cloud of Unknowing*, Patrick J Gallacher ed., Kalamazoo, Michigan: Medieval Institute Publications, 1997.

Kharms, Daniil, *Incidences*, Neil Cornwell trans., London: Serpents Tail, 1993.

Libet, Benjamin, "Time of conscious intention to act in relation to onset of cerebral activity (readiness-potential): the unconscious initiation of a freely voluntary act", *Brain* 106, 1983, pp. 623–642.

Lucretius, "On The Nature of Things" (translated as On The Nature of the Universe), James H Mantinband trans., New York: Frederick Ungar, 1965.

Padmasambhava, *The Tibetan Book of the Dead*, Gyurme Dorje trans., London: Penguin, 2005.

Parmenides, "On Nature" fragment 6, translation adapted from those of John Burnet, Richard McKirahan and Arnold Hermann.

Soon, Chun Siong, Marcel Brass, Hans-Jochen Heinze, John-Dylan Haynes, "Unconscious determinants of free decisions in the human brain", *Nature Neuroscience* 11, 2008, pp. 543–545.

NULL AND OBJECTIVE: LONDON FIELDWORKS AND METZGER'S BRAIN

Nick Lambert

Neither the art world nor society has woken up to the fact the artist engaged in technological art is performing a role which is new in a quantitative and qualitative sense.[1]

Null Object combines Bruce Gilchrist and Jo Joelson's exploration of interfaces to directly capture the brain-states of their subjects using electroencephalograms (EEGs), with the direct contribution of Gustav Metzger's own "thoughts of nothing". Metzger's involvement with the project might be seen as purely informational, in the sense that he was contributing his EEG readings to construct the 3D shape that would then be removed from the cube. However the whole project should be understood in terms of Metzger's lifelong entanglement with technology in the arts, his attempts to comprehend technology's meaning within an arts context, and London Fieldworks' projects to examine emotional states and the environment.

The collaboration between London Fieldworks and Metzger might be seen as fortuitous, as they all live and work in a small artists' community in Hackney. Yet there is also a clear parallel between the approach that London Fieldworks were already taking before their contact with Metzger, and his understanding of the artist's role in exploring the sciences that he developed during the 1960s.

These aspects came to the fore in *Syzygy*, 1999, a collaborative work conceived by Gilchrist and Joelson, performed on the island of Sanda in the Western Isles of Scotland. It explored the analogy of emotional states and the weather, using kites flown at altitudes of 1000 feet to measure the wind speed, temperature and light whilst biofeedback sensors monitored the kite-flyers' brain activity using EEG monitors. A team of writers, programmers and musicians responded to the kite flying, and repurposed the weather data for

Syzygy
Sanda Island, Scotland, 1999
Photography Anthony Oliver

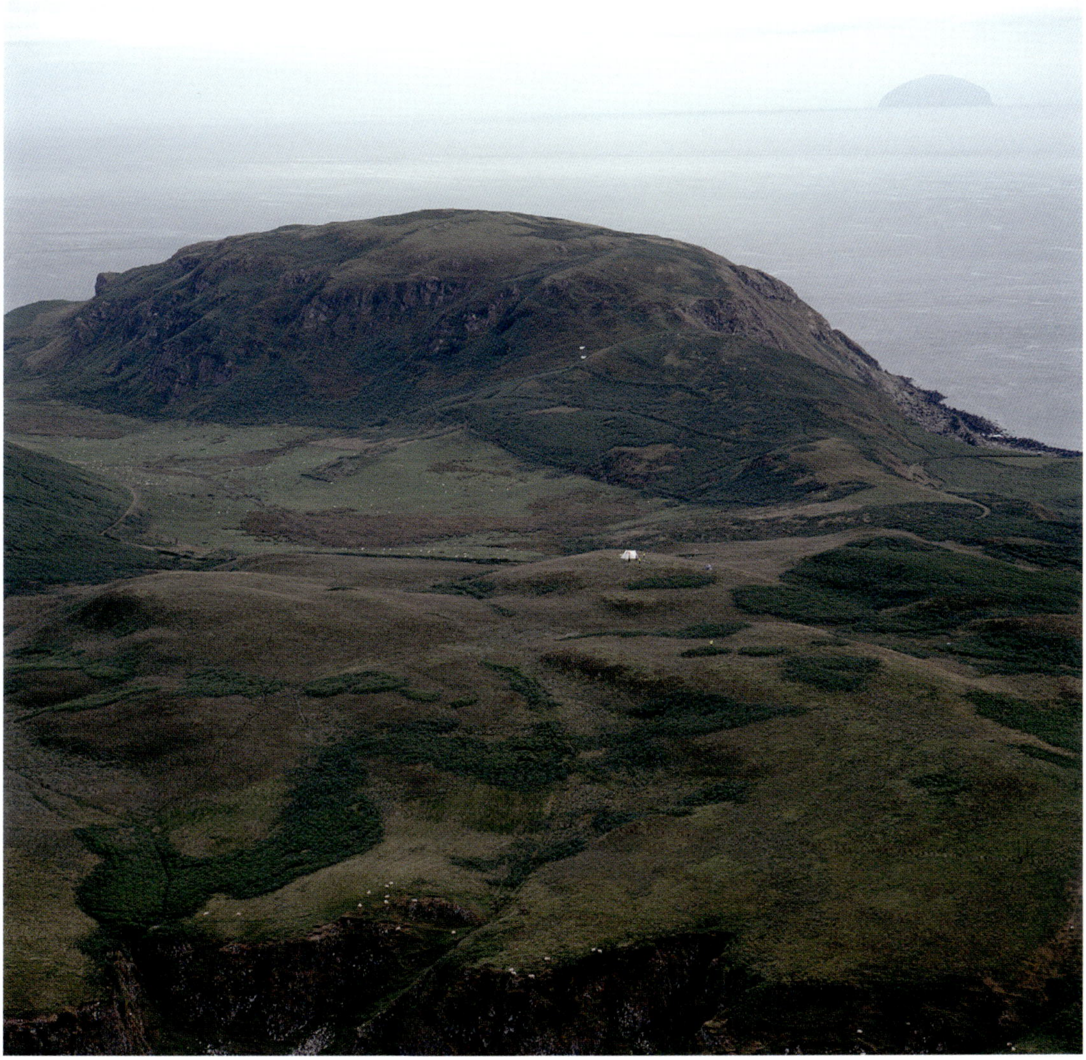

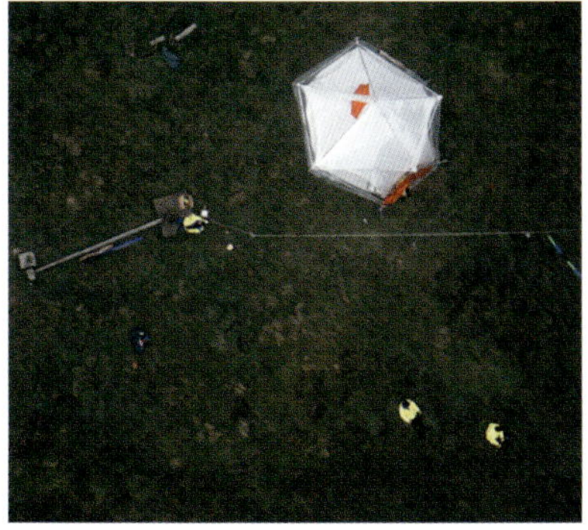

their own satellite projects, and the results were communicated back to a telematic sculpture at the Institute of Contemporary Arts (ICA) in London:

> Simultaneous weather and brain data was metaphorically correlated, by its transmission and manifestation in a 'smart' sculpture 700 miles away at the ICA in London. This sculpture was a glass bowl suspended on a steel spine, encrusted with light emitting diodes and housed in an electro-reactive glass tank. The glass bowl looked like a large cranium, being a little larger than head size.[2]

In this way, *Syzygy* moved beyond a direct correlation of scientific measuring and artistic response, by integrating several quite different manifestations of the work, including a data-driven sculpture and the project website. It did not embody any particular scientific theory but rather deployed monitoring techniques to gather the data that then became part of the project's raw material. It was facilitated by contemporary communications technologies but was essentially about the interrelation of techniques and human responses "through poetic application". [*Syzygy* website] This theme is picked up by Oliver Bennett:

> [As] the sparkly pendant dangled from the ceiling [of the ICA], it seemed to harbour the very emotional crux of new technology— the glamour of progress offset by the fear of sentient machinery. It is also telling how projects such as *Syzygy* bring to mind the archaic language of animal control—the human endeavour to 'harness' the wilderness and in this case, the wildest of manifestations; weather itself.[3]

Thus the informational exchange of sampled weather data, plus bodily responses from physical participation combined with the musical and textual interpretations of the artists were broadcast via the website and delivered to an audience physically located in the heart of London, from a remote island at the margins of the UK. In developing and delivering *Syzygy*, Gilchrist and Joelson were advancing the scope of artistic engagement with communications technologies and, arguably, the conceptual and experiential range of art as well. This chimes well with a statement made by Metzger at the conference and exhibition New Tendencies 4, concerning computer technology and art in 1969:

> Through his theories and work the artist integrates information on new techniques in science and technology; he functions as a processor and distributor of information.[4]

However, as Metzger himself states in his foreword to the publication *Little Earth*, concerning the eponymous art project by London Fieldworks in 2005, it is the ethical engagement with science through art that is of great importance; and in *Little Earth* this emerged most strongly in the interplay between landscape, location, the scientific pursuit of data in all conditions and climates, and the human element at the heart of supposedly objective scientific work.

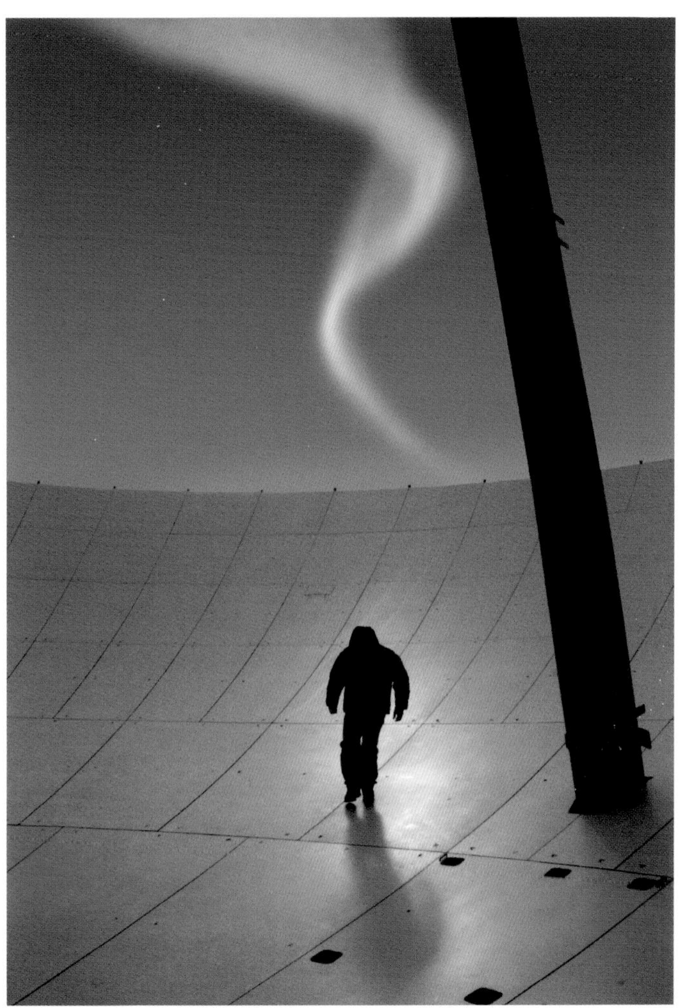

London Fieldworks on location during
the making of *Little Earth*, 2004
Fixed 42 m UHF parabolic antenna,
Eiscat Svalbard Radar, Frontiers 8

As Tracey Warr says:

> A lot of current 'sci-art' [is] still mired in the idea that art can be
> an illustrative, accessible, user-friendly mediation for science, or
> that science offers art an alluring range of kit and language to be
> appropriated.... Gilchrist and Joelson's work bypasses the binary
> fallacy of sci-art. [It] approaches a complex knot of belief, desire,
> creativity and knowledge.[5]

In his comprehensive overview of science-art collaborations,
Information Arts, the late Stephen Wilson drew together numerous
strands of practice that engaged with advanced technology and
scientific institutions, or individual artists and scientists. Although

artists and scientists had collaborated informally for decades, it was only after the advent of Experiments in Art Technology (EAT), following the groundbreaking Nine Evenings exhibition at the New York Armory in 1966, that 'Art-Science' really gained for itself a specific niche, rather than being an occasional crossing of disciplinary boundaries.

At the time, this was against the backdrop of CP Snow's stern warning in "The Two Cultures" that the two poles of Western culture, the sciences and the humanities, were pulling ever further apart across a chasm of mutually incomprehensible thought. EAT was not perhaps a direct response to this—it arose from its founder Billy Klüver's conviction that artists should be given access to the latest advances—but it certainly set the tone for the collaborative ventures of the late 1960s, not least Cybernetic Serendipity, organised by Jasia Reichardt at the ICA in 1968, and Software by Jack Burnham in 1970. The two shows were almost opposite in their aims, scope and success, the first generating much optimism about the possible combined efforts of art and computers, the second near-undermining them and leading to the eventual abandonment of Burnham's cybernetic-influenced aesthetic.

The 1970s were to lead to a much more critical engagement with aspects of technology, in particular; partly through increased awareness of environmental destruction and partly through the emergence of new technologies in art, especially video which eventually found a place in the mainstream. However, arts-science continued in various areas, supported and documented by *Leonardo* journal whose founder, Frank Malina, wanted to provide a forum for this convergent area, and enable its development. In this context, the growth of telematic and interactive art, using new communications interfaces that would eventually be named the "Internet" but already existed by the late 1970s, would be an important factor. The idea of interfacing the human and the machine to produce new kinds of art was as old as the early twentieth century, but with the rise of new sensory input apparatus, new forms of interaction could be found.

It is in this context, perhaps, that *Information Arts* provides the best overall picture of this emerging area: as the continuing exploration of interface and response mechanisms that are quite different from the visual/tactile Human Computer Interface that has come to characterise modern digital systems. As Wilson said in his first chapter:

Art that explores technological and scientific frontiers is an act of relevance not only to a highbrow niche in a segregated corner of our culture. Like research, it asks questions about the possibilities and implications of technological innovation. It often explores different inquiry pathways, conceptual frameworks, and cultural associations than those investigated by scientists and engineers.[6]

This shared parallel process of enquiry, though with quite different motives, is perhaps the "interface" (in the more abstract sense) that brings scientists and artists together to engage in combined projects. It can enable the artist to benefit not only from uncommon techniques, that are just being prototyped, but also ways of approaching new concepts and forms of enquiry that originate in a different worldview. This is not to over-emphasise the underlying differences between the arts and sciences, otherwise there would be no possibility of convergence, but rather to show that the conceptual interchange can happen at several levels, and it is not solely about technique. Indeed, Wilson cautions that mere deconstruction of the scientific process—though implicit in some of the critiques of technology that appear in arts-science work— can act as a barrier:

> However, in its rush to deconstruct scientific research and technological innovation as the manifestation of metanarratives, critical theory leaves little room for the appearance of genuine innovation or the creation of new possibilities. While it has become predominant in the arts, it is not so well accepted in the worlds of science and technology.[7]

Thus a different fault-line has appeared: not so much in the acceptance or rejection of science in the arts, but an argument over its significance and interpretation. The critical viewpoint predominantly sees technology as problematic—though perhaps overestimating its power—whilst the prevailing scientific/ technological viewpoint engages with the creativity of development and looks more towards ongoing future developments and their extensions of potential. Both worldviews are of course unlikely to abandon their basic premises because each is entrenched in its own institutional culture, yet cross-platform art projects must necessarily involve a degree of interchange between these cultural territories.

Gustav Metzger seems to have embodied both aspects in a remarkably consistent way throughout his career. Perhaps because he drew his main creative inspiration from his concept of "Auto-destructive art" from an early stage, he was able to embrace the more unusual propensities of a technological society in terms of process and information, whilst still being highly aware of its military origins and basis. Whilst he did not subscribe to a somewhat naïve technological messianism that certain 1960s artists did—even Burnham to some degree—neither did he renounce technology in a reactionary sense and refuse to sully his hands with modernity. Rather he saw potentials that if fully exploited could expose the darker or destructive aspects and yet embrace the evolving, almost lifelike, side of the computationally-based society that emerged after the Second World War.

In 1969 the organisers of the Computer 70 exhibition asked the Computer Arts Society to design a central feature for the exhibition, which was held at Olympia in 1970. The result was a pioneering multiplayer interactive simulation of resource exploitation and ecological results: the *Ecogame*. It was housed in a dome with real-time responses from computer terminals and computer-controlled slide projectors. It was coordinated by CAS founders John Lansdown and George Mallen, and informed by the latter's work with cybernetics.
Illustration courtesy
George Mallen, c. 1970

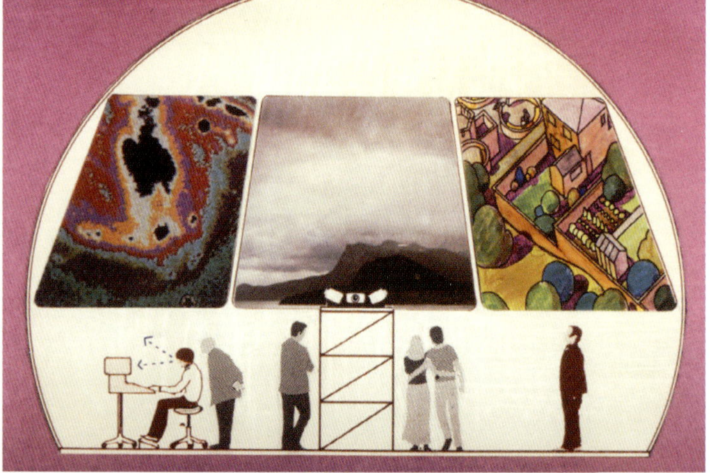

Metzger was notably engaged in one of the key British responses to the inherent possibilities of digitally-mediated art, the Computer Arts Society, founded in 1969 following a meeting at the 1968 IFIP conference between programmer and musician Alan Sutcliffe, cybernetician George Mallen and architect John Lansdown. Each brought a particular vision of a way to use computers in the arts—which were always in the plural for CAS, and not merely visual—and were strongly influenced by theories about the informational nature of aesthetic appreciation. CAS provided a forum for an emerging area that otherwise had few outlets, being too new for the mainstream art world and yet also outside the boundaries of

EVENT ONE
COMPUTER ARTS SOCIETY

Front cover Event One catalogue,
Computer Arts Society, 1969
Image courtesy Computer
Arts Society

computer science. Many of its members held dual roles as computer engineers and artists—though others were indeed full-time artists such as Malcolm Le Grice—but the implications of such boundary crossing were revolutionary at the time. This was also the era of the Drury Lane Arts Lab, later IRAT (Institute for Research in Art and Technology) under John Lifton, and the Artists' Placement Group that sought to engage artists with companies.

Metzger became involved with CAS in late 1968. According to his recollections, a chance conversation with someone in the street led to a connection with this arts-computers group even before its official emergence. In this respect Metzger would seem to be incredibly lucky, though the role of synchronicity should not be overlooked.

At its inaugural exhibition, Event One, at the Royal College of Art in April 1969, Metzger proposed a work entitled *Five Screens for Computer* and produced an initial drawing and model (see image overleaf). It would consist of five steel frames, 30 feet high by 40 feet long (producing the 4:3 ratio of a standard screen) containing at least 1200 cuboid elements that could be advanced or retracted within the frame like physical pixels, before being ejected up to 30 feet by mechanisms within the frame. This ejection would be partially pre-determined by a computer program that controlled the graphic output of the piece, and partly by random environmental occurrences such as intense sunlight. The line-drawing in the image shows a frontal view of one screen with some elements ejected, creating a hollow at the centre.

Thus began a partnership that would last until the middle of 1972, when increased demands for work in his other career as a valuer of antique books caused Metzger to finish as editor of CAS. During that time, the direction of the journal *PAGE*, not least in requesting from *PAGE* 4 onwards that different people should design each issue, leading to a profusion of interesting forms that make the look very distinctive, even today. Metzger also tried to

FIVE SCREENS WITH COMPUTER (1963-69)

GUSTAV METZGER

This auto-destructive sculpture is to be made
of stainless steel. It consists of 5 frames
40ft. long, 30 ft. high, and 2 ft. deep. These
frames will be spaced 30 ft. apart-staggered in
plan. Each frame will be packed with at least
1200 uniform elements 2 ft. in length, with a
square or rectangular face. These elements can
be moved forwards or backwards within a frame
at controlled speeds, and will finally be ejected
at various controlled speeds, reaching a maximum
distance of 30 ft. After a period of ten years
in which all the elements will have been ejected
and the frame disintegrated in stages, the site
will be given over to another use.

The computer is used in three areas.
DESIGN OPERATION RECORDING

DESIGN

Since all the decisions on the activity of the
screens will be made before production begins,
it is necessary to have the most complete
understanding of the work's potential at the
design stage. A computer allied to graphic
output will be used to plot the numerous
possibilities for moving and ejecting elements,
and for visualising the possible shapes of
the screens in transformation.

55% of the elements will be ejected on a
pre-determined programme. The rest (including
one entire screen) will be ejected in a random
manner. These random ejections will be sparked
off by intense sun or electric light, or by the
assembly of people above a certain number in the
vicinity of a screen. Random ejections are
subject to a variety of controls such as structural
considerations, and will be co-ordinated with the
overall programme.

OPERATING

A computer will be in general control of the
electro/mechanical activity of the sculpture -
continuous adjustments (on-line) will be
necessary. The computer will also direct peripheral
activity such as the raising of the glass wall
surrounding the site before ejections can take place.

RECORDING

The computer will be used to print out and draw
the day-by-day development of the screens. This
will be necessary to check on operational, structural, and safety factors, and will be an aid to
maintenance activities. This graphic output, along
with photographs and films, will be preserved as
part of the documentation on the work.

This drawing is the last in a series of ten showing the
development of one screen (No. 3) in the first three years
of its activity. It was produced by Mr. D.E. Evans,
Computer Unit, Imperial College, London in March 1969,
on an IBM 7094 11 (32K memory), with CALCOMP plotter.

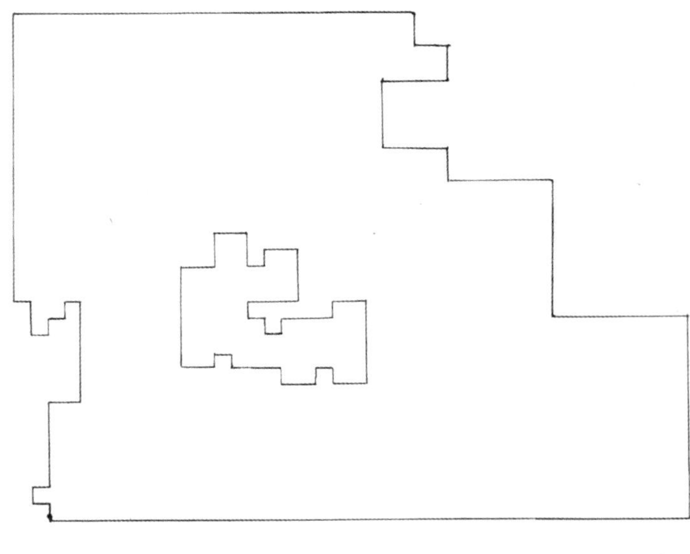

Gustav Metzger, *Five Screens for Computer*, 1969, proposal in Event One catalogue, p. 18, Computer Arts Society, 1969 Image courtesy Computer Arts Society

instil—with some success—the values of the avant-garde into the new publication and make it as wide-ranging as possible, acknowledging both the advances in computing technology and potential artistic uses, but also their burdensome origins in a scientifically-directed military complex that was far from neutral in its quest for dominance of all spheres. The Internet, after all, began life as ARPA Net.

Although many modern commentators assert that the 1960s art-technologists were naïve in their optimism, it is closer to the truth to say that many chose to be optimistic despite knowing the choices and contradictions involved in technological art. Metzger, being even more aware than most and already immersed in a critical worldview, was not such an optimist but he was highly willing to engage and even shape the nascent computer art movement with his particular view of the technological zeitgeist. As he said in his "post mortem" of the Event One exhibition in 1969:

> For artists or their organisations to approach science and technology as if they were entirely desirable and constructive forms of activity is extremely unrealistic and dangerous. There is now a clearly established and growing trend amongst [some] scientists and technicians [to] expose and correct the disastrous alliance of science and technology with exploitative and destructive social systems.[8]

Metzger's move towards computer art in 1968 was not simply coincidental, but entirely in keeping with his future-focused vision of the computer's critical role in society, a role he hoped to divert in a more socially-responsible and arts-oriented direction. When he finally stopped editing *PAGE*, he had perhaps achieved as much as he could within the CAS structure and moved away from engagements with this type of digital art, at least until he started working with London Fieldworks.

In the cases of both Metzger and Gilchrist and Joelson, the integration of scientific methods or techniques is not what is often referred to as "sci-art" or "art-science". Gilchrist contrasts the often-didactic connotations of those terms with the primarily conceptual thrust of London Fieldworks, examining the role of the individual in relation to a technologically mediated culture and balancing that against nature or the "natural", insofar as one can find this in contemporary life.[9]

In terms of *Null Object*, the project builds on the data that Gilchrist has been collecting for some 13 years from psychological

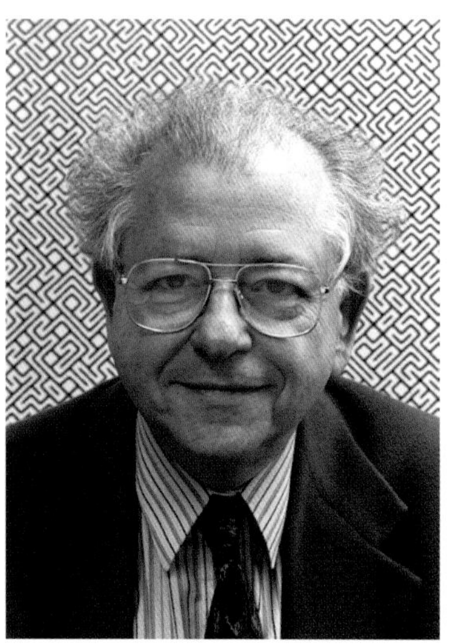

Béla Julesz, in front of a picture from his and A Michael Noll's computer art exhibition, Computer-Generated Pictures, held at the Howard Wise Gallery, New York City, in 1965. Photograph courtesy Rutgers University

experiments about the perception of images and brain response. It has its origins in the digital live-art installation *Divided by Resistance* by Gilchrist and Jonny Bradley that won first prize from the ICA/Toshiba Art & Innovation commission in 1996. It involved Gilchrist, over a one-year period in his studio, sleeping wired up to an EEG and using infrared eye tracking to wake him after a predetermined time after the onset of Rapid Eye Movements (REM) so this activity could be EEG recorded. The audience interacted with him whilst he was asleep in the gallery during the performance, inputting mild electrical currents in the form of a crude morse code he had trained himself to recognise.[10] It was further developed for the *Thought Conductor* performances and then for a project called *Looking at Primitives*. Taking as its basis the geometric primitives, and seeking to understand how audiences commodify and objectify art, the project developed a "perception depository" of EEG readings from participants looking at single image stereograms. The participants would record their perceptual processes at the very moment when the image—a primitive solid—emerged from the random dots of the stereogram. The resulting EEG trace shows the burst of brain activity at this moment of recognition.

There is another connection here with the early years of computer art. The Hungarian-American psychologist of vision Béla Julesz worked with the pioneer computer artist and graphics specialist Michael Noll. Their first computer images were exhibited at the Howard Wise Gallery in 1965 as part of the pioneering computer art show Computer-Generated Pictures. These images formed the basis for the random-dot autostereograms, developed by Christopher Tyler, a student of Julesz, shown to participants in Gilchrist's work.[11]

The database records the waveforms of the participants and relates this to the 3D primitives they were viewing via the stereograms. The key element in *Null Object* was the connection with Metzger's thought processes in the consideration of "nothing", then connecting these to the form generated by his EEG output. As Gilchrist and Joelson said in their initial proposal for the project:

> In *NULL OBJECT*, attempting to empty the mind of impressions will create an interactive residue of 3D shape information from the Perception Depository, leading to the production of an artwork via industrial process out of a kind of distributed authorship.[12]

It was this output that was translated into commands for the industrial robot used for stonecutting, thanks to Yaroslav Tenzer's work at Imperial College London and latterly at Harvard. The particular challenge for the roboticists is that under normal circumstances, the robot is subtracting (cutting) material from the surface of a given material, whereas here the cutting has to be done inside the volume of the object. The eventual solution was to slice the block in half, hollow out each section, then reunite the slabs as closely as possible.

There is a key question about the role of the computer as a mediator between the interior world of the participants, at least as represented by the brain responses to the stereograms, and the interior world envisioned by Metzger when he thought of "nothing". By making the electrical impulses equivalent to images through digital processing and storing them in the database, the computer provides a way of translating from one person's brain to another without using the usual human tools of language or drawing to convey these states. The final form from which the 3D model is produced for the robot sculptor derives entirely from Gustav's EEG responses.

Thus a form of long-distance collaboration occurs, unusual in Metzger's oeuvre where few if any of his works do not feature him as the sole agent. But here his passivity is an essential part of the piece and his withdrawal, which is a "negative act", is at its centre. The "image" in the void can be seen as a kind of a residue from Gustav's struggle to think about nothing. This resembles a Buddhist viewpoint on the need for liberation from the material world, but it is more informed by Metzger's background, and his materialist dialectic, rather than anything mystical as such. It could be regarded as the fallout from trying to achieve this state of mind.

On the surface at least, he seems remarkably nonchalant about his role. When we met, Metzger stated that he had come into the project on the basis that he would "stay out" and let it develop in its own way once he had made his negative contribution. However this was perhaps belied by his fascination with the material record of *Null Object*, in the form of a test printing of the cube that was about two inches square.

Seeing the printed model for the first time with the opening revealing the hollowed-out interior like the inside of a geode, Gustav observed, "It's like the *Venus of Willendorf*". He seemed to be referring to the shape of the entrance, which did indeed resemble the outline of the famous paleolithic statue. Clearly there was also some link between the form and caves, and memory, and the discovery of sculpture by early humans. Although this was a spontaneous pronouncement, it seemed that Metzger had in fact been thinking about the implications of *Null Object* and was still involved in it on several levels.

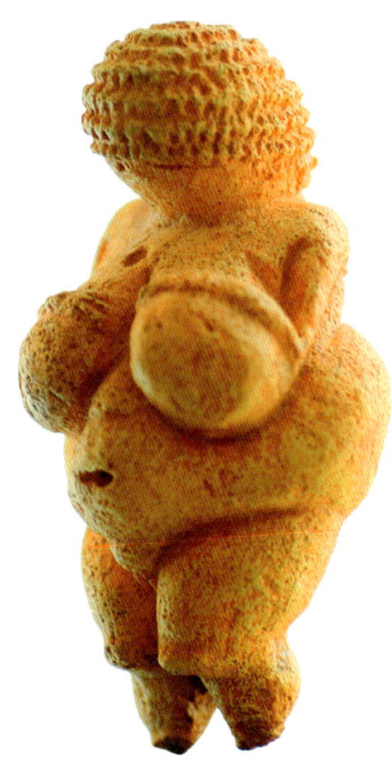

Venus of Willendorf
Photography Matthias Kabel, 2007

Nonetheless, his direct involvement with the production of the object itself was limited to the role of delivering his passive thoughts to an observant machine, capable of recording and interpreting them. This use of interfaces reflects several other projects by London Fieldworks over the past 15 years and shows how technology can be employed as mediator, so that the finished object will have no human involvement in it, not even any finishing by hand. The path from the brainwave to the stone is entirely mediated by technology, apart from the final reassembly of the two halves of the stone, rather like brain hemispheres.

Null Object also departs from most of Metzger's earlier works in terms of its collaborative aspects and the fact that it creates a final solid, presumably lasting object. For all that it embodies the responses of the participants of *Looking at Primitives*, and Metzger's own "thoughts on nothing", it is reified in a material form that places it in a line of descent from the *Venus of Willendorf* and all its earlier antecedents. That is not to say that *Null Object* consists only of, or is contingent upon its outcome in stone—it is part of a complex of immaterial and responsive elements, and very highly developed—but its concrete existence is most interesting. Whilst the preservation of what is broadly termed "media art" is very much an emerging area, the carved stone itself will be lasting testimony to the project.

In this sense, one can appreciate a slightly different understanding of the interplay of science, technology and art that is articulated by Brian Winston in his paper, "A Mirror for Brunelleschi". He contrasts the positions of Walter Benjamin's view of History as catastrophe, viewed through his interpretation of Klee's painting *Angelus Novus*, and the more optimistic views of technological progress. He then looks at the possibility that both pessimists and optimists in this area are overly deterministic, and perhaps over-stress the mastery of the machine. He sees technological development, and its resultant incorporation in the arts, as a much slower process and evidence turns to the first demonstration of single-point perspective by Brunelleschi. He says:

> There are two views that one can take of technological progress in our contemporary culture. One can assume that technology, like History in Benjamin's description, is a catastrophe; or one can believe that it is an ever-growing pile of discrete and wonderful events moving society toward some sort of utopia. Of course, lesser variants on either of these positions are possible. Cognitive dissonance allows one to hold to both positions simultaneously.[13]

Looking back at London Fieldworks' oeuvre, beginning as far back as *Divided by Resistance*, one sees the technology in a facilitative role and the science as a focus, though not perhaps the subject of the work as such. It is also about interfaces, but these constitute an aspect of the work—even pieces like *Syzygy* that required long-distance networks to be able to function at all. These pieces simply could not have existed before the development of digital communications, at least not in their current form. But unlike other artists, the medium of communication and the Net is not the subject either. The focus is perhaps on the emotional and informational interchange that can happen across these distances, and especially the interchange of physical states that only specialist equipment like the EEG can measure.

Thus in these works, the scientist and artist both inhabit a world based on measurement, but the use to which the information is put is quite different. Gilchrist and Joelson draw on specific technological and scientific experience, in the service of a concept that has some metaphorical link to the underlying science without simply rendering it in artistic form. This is a transformative process, much like the link between the science of sound to the construction of musical instruments and ultimately the musical performance itself. The scientific element of music is not necessarily made manifest, except perhaps in some of its more formal structures or experiments, but it underpins the sound and quality of the playing and also makes possible the complex harmonics of large orchestral pieces.

London Fieldworks might be seen to some extent as performers with data, in the sense that their works involve the collection, coordination and construction of datasets that are then rendered in a variety of forms. In Steve Dixon's large opus *Digital Performance*, he considers several approaches to this performative aspect, drawing on other theorists' definitions:

> In relation to digital arts and video installations, Margaret Morse suggests the interactive user takes on "the virtual role of 'artist/installer' if not the role of artist and inventor of that world". Bolter and Gromala offer the substitution of the word "performance" as "an even better word than interaction to describe the significance of digital design in general…".[14]

Yet *Null Object* is not, in its final rendition, an interactive work as such; it presents a range of sensory experiences including that of investigating the stone block with one's hands, but the work exists in a halo of interrelated processes. It shows, I think, the range of possibilities that are subsumed under the title of "creativity", be it the creative understanding of the technological sphere when it is put to uses not solely utilitarian, and the artistic sphere when it is informed by, but not controlled by, scientific theories of perception and response. Might there be some shared creative ground within which the sciences and arts can at least cohabit, if not completely mingle and produce a neither-nor hybrid? London Fieldworks' concepts are considerably most subtle in this regard. By investigating the creative aspects of scientific processes, they subvert not only a purely scientistic worldview, but also the attitude that science is purely about quantifying, measuring or controlling. In this they seem to concur with David Bohm, the distinguished physicist of Birkbeck College, who wrote:

> Scientists are seeking something that is much more significant to them than pleasure. One aspect of what this something might be can be indicated by noting that the search is ultimately aimed at the discovery of something new that had previously been unknown. But of course, it is not merely the novel experience of working on something different and out of the ordinary that the scientist wants…. Rather, what he is really seeking is to learn something new that has a certain fundamental kind of significance: i.e. a hitherto unknown lawfulness in the order of nature, which exhibits unity in a broad range of phenomena. Thus, he wishes to find in the reality in which he lives a certain oneness and totality, or wholeness, constituting a kind of harmony that is felt to be beautiful.[15]

It is for the artist working around, and through, various scientific paradigms to comprehend, construct and subvert them; not perhaps to paint science in a negative light as with certain kinds of protest art, but rather to understand it in terms of its cultural impact and import. In 1968, just such a possibility was articulated by Gustav Metzger and it undoubtedly informs his own interaction with science and technology:

> It is a sad mistake [to] envisage the artist as a parasite, exploiting the technical advances of others. From experience, I can say that quite often, the artist wants to do things that are more advanced than existing techniques permit.[16]

NOTES

[1] Metzger, Gustav, "Tendencies 4", 1969, quoted in *A Little-Known Story about a Movement, a Magazine, and the Computer's Arrival in Art*, Margit Rosen ed., Cambridge, MA: MIT Press/ZKM, 2010, p. 423.

[2] Warr, Tracey, "Tuning In", *Syzygy/Polaria*, Bruce Gilchrist, Joelson, Jo, eds., London: Black Dog, 2001, p. 6. Also at www.londonfieldworks.com/projects/syzygy/about.php

[3] Bennett, Oliver, "Kites: An Aerial Review", *Syzygy/Polaria*, p. 19.

[4] *A Little-Known Story about a Movement, a Magazine, and the Computer's Arrival in Art*, p. 423

[5] Warr, Tracey, "Measuring Beauty in the Upper Ice World", *Little Earth*, Bruce Gilchrist, Joelson, Jo, eds., London: Black Dog, 2005, p. 17.

[6] Wilson, Stephen, *Information Arts*, Cambridge, MA: MIT Press, 2003, p. 3.

[7] Wilson, Stephen, *Information Arts*, p. 11.

[8] Metzger, Gustav, "Notes on the Crisis in Technological Art", 1969, pp. 2–3, typescript handed out at "Post Mortem on Event One", British Computer Society, 3 April 1969, in London.

[9] From a discussion with Bruce Gilchrist at the Digital Weekend, Victoria & Albert Museum, Sept 21 2012.

[10] Walwin, Jeni, *Low Tide: Writings on Artists' Collaborations*, London: Black Dog Publishing, 1997, p. 11. Also online at www.artemergent.org.uk/dbr/dbr.html

[11] See Gilchrist, Bruce and Jo Joelson, "Looking at Primitives", online at www.artemergent.org.uk/lap/lap_1.html

[12] Gilchrist, Bruce and Jo Joelson, Project Proposal for *Null Object*, October 2011.

[13] Winston, Brian, "A Mirror for Brunelleschi", *Daedalus*, vol. 116, no. 3, Futures, Summer, 1987, p. 187.

[14] Dixon, Steve, *Digital Performance*, Cambridge, MA: MIT Press, 2007, p. 560.

[15] Bohm, David, "On Creativity", *Leonardo*, vol. 1, no. 2, April 1968, p. 138.

[16] Metzger, Gustav in *A Little-Known Story about a Movement, a Magazine, and the Computer's Arrival in Art*, p. 425.

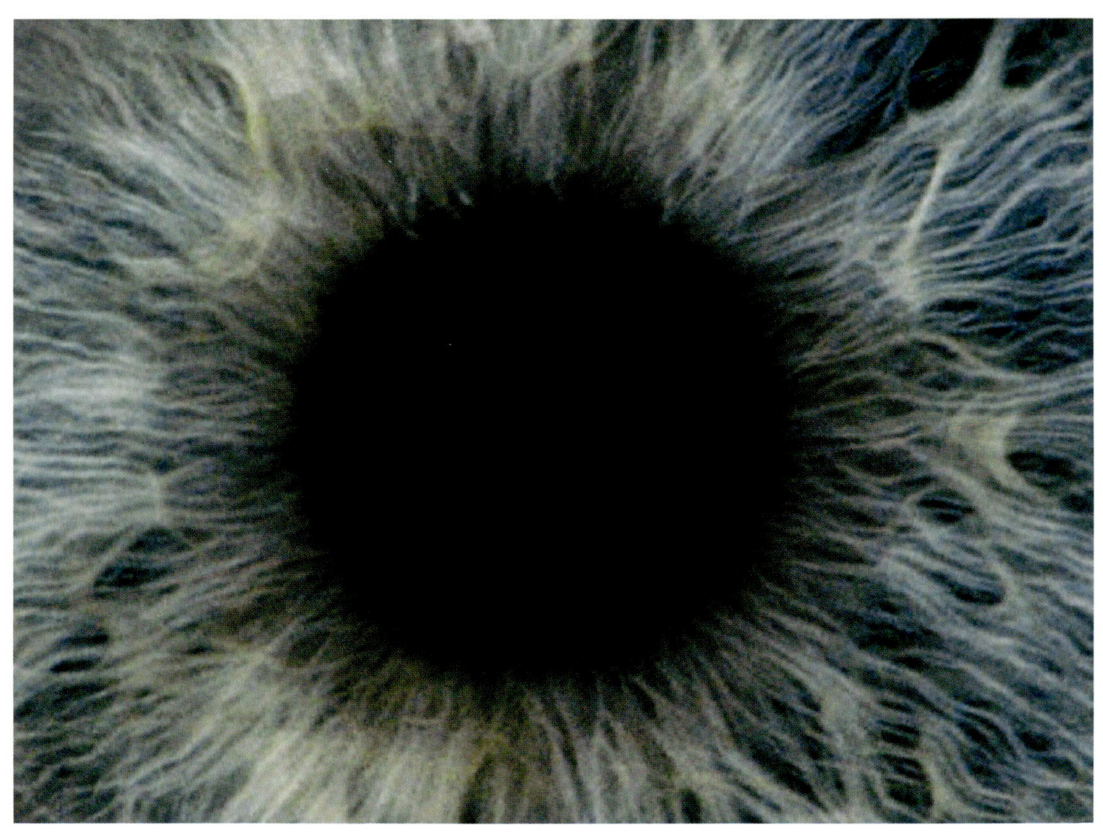

Close-up of the human pupil
Modified from Look into My
Eyes, 2012

NULL OBJECT: PROLEGOMENA

Christopher W Tyler

The core goal of the *Null Object* project is conceptualise the inconceivable—what it means to think about the absence of any object, the lack of an object, the non-existence of an object, and so on. In principle, the concept of nullity may be addressed in many forms, one being the philosophy of the 'void', which is a concept with an illustrious history. In early Greek philosophy, it finds expression as the aperion of Anaximander (the 'limitless' or 'inexpressible' essence from which all things derive and to which they all return). The aperion is understood to be a similar concept to Chaos, the original state of the 'gaping void' or vacuum that nevertheless incorporates energy from which the universe originates. Since he never clearly defines the term, however, saying that the essence of all things is the aperion amounts to a tautology, that the essence of all things is the universal essence. This leaves us with the conceptual void that we are staring into an answer that has no meaning.

The Roman poet-scientist Lucretius finds hundred of lines to say about the void in his poem "De Rerum Natura" (On the Nature of Things, c. 50 BC), and modern quantum mechanics maintains that space in the absence of all objects has an infinite energy level in order to derive their equation structure, a term developed by Einstein and called the "null point energy". The infinite value of the energy at every point in empty space, which relates to the "null space" of the present project, is just one of the many infinities encountered in the exposition of quantum theory. This mind-bending confrontation between nothing and infinity seems, at its core, to be ultimately as vacuous as Anaximander's concept of the aperion. If a theory of essence encounters such an array of infinities, one is led to ask whether it is based on a fundamental misconception in its basic formulation, as successful as it has undoubtedly been in accounting for the primary observations for which it was designed.

It may be conceptualised ontogenetically, in the sense of the thought processes of a mind that has never encountered any objects because it was formed in a domain lacking any things that could be described as "objects". For humans, this would be a practical impossibility, but it might be possible to envisage some kind of sea creature that developed in the open ocean and drew nutrition from the liquid medium without ever encountering any objects. This might be an approximate description of the life cycle of some kinds of jellyfish, but they do not have much of a 'mind' in which to evaluate their potential thought processes. Indeed, the very primitiveness of their neural net approximation to a primitive brain is suggestive of the minimality of the thought processes required in the ontogenetic absence of an environment of objects. Since the nutrition is, by definition, not segregated into discrete elements that would conform to the definition of objects, very little structured representation is required for the organism to access it (nutrition being, of course, the primary need for any organism, along with self-protection and procreation).

In the vein of minimalism, it is also relevant to point out that a mind as usually conceived is not, in fact, a prerequisite for the kind of interaction with objects implied by these nutritional considerations. It might be thought that the requirements to hunt for, identify and ingest objects as food would necessarily require the processes of mental representation describable as 'object perception'. In this connection, it is salutary to point out that this full suite of activities is engaged in by single-celled organisms that, by definition, do not have a brain. The amazing planktonic creatures known as dinoflagellates perform all these activities, despite being only a single cell. They are equipped with both a flagellum, or whip-like tail, that allows them to locomote rapidly through the water, and a light-sensitive eye-spot with a lens above that provide a primitive imaging capability. Dinoflagellates may be observed to hunt moving cells as prey, and to track particular cells through complex environment of other objects, ingesting them when caught, all without the benefit of a brain. Although they have elaborate calcite skeletons of astounding complexity (and symmetry), they are single-celled organisms. They must, in fact, have some form of external guidance system, presumably implemented by the genetic control of the dynamic microtubular cytoskeleton common to all single cells, but the operation of such processes as an microminiature analogue of neural processing remains a speculative domain of neurobiological conceptualisation.

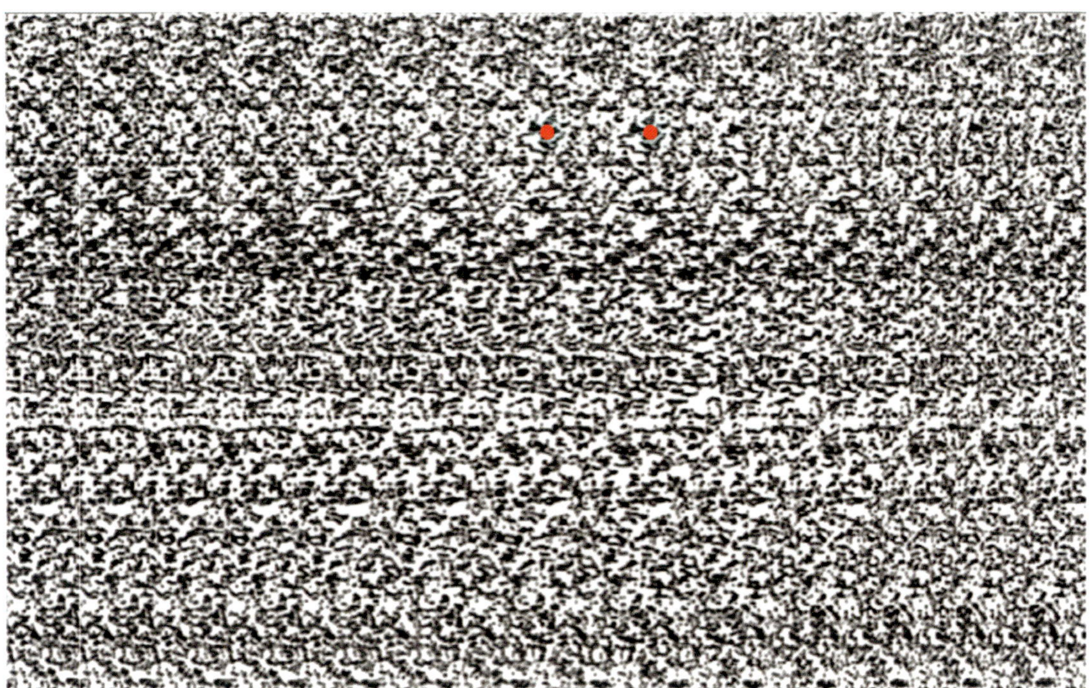

Random-Dot Autostereogram
1990
Image courtesy Christopher Tyler
Copyright Christopher Tyler

In terms of the null object representation in the human brain, we may approach it via a confrontation between three disjunctive concepts: the scientific representation of negative 3D structure in perception; the inverse representations of negative space in art praxis; and the inchoate representation of the null object, which in principle has no form, 3D structure or content. With respect to 3D structure, it should be recognised that an object is an inherently 3D concept. It is helpful to consider Webster's definition of an object as "a discrete tangible or visible thing". Thus, the essence of an object is its discreteness, its aggregation into a connected agglomeration, forming a 'thing' as opposed to a distributed material that could be termed 'stuff' (Adelson, 2001). I would argue that a primary property of 'thinghood' is also its three-dimensionality, its essential ability to be viewed from various angles and otherwise manipulated in relation to the 3D world. Visual images may be a proxy for the 3D objects, but they are rarely viewed without perceiving some level of 3D interpretation of the implied structure.

One form of null object is therefore obtained through the perceptual representations of 3D structure implemented in a form abstracted from its 2D cues, which can be achieved by the use of the cyclopean random-dot stereogram stimuli developed by

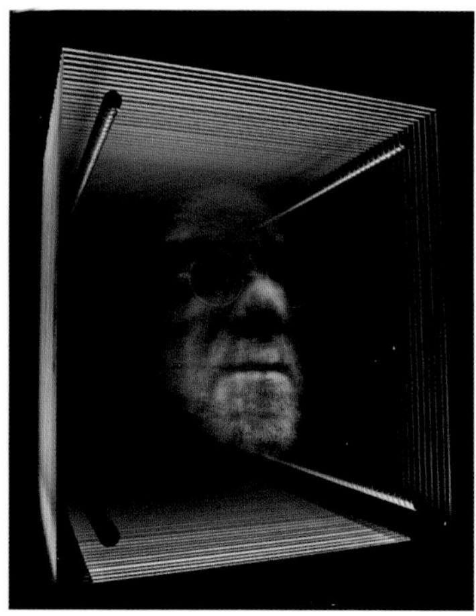

The Nature of the Beast 4: Chuck
NFN Kalyan, 2011
Glass, wood and steel
27 x 25 x 26 in
Photography Pipo Bonamino
Courtesy the artist

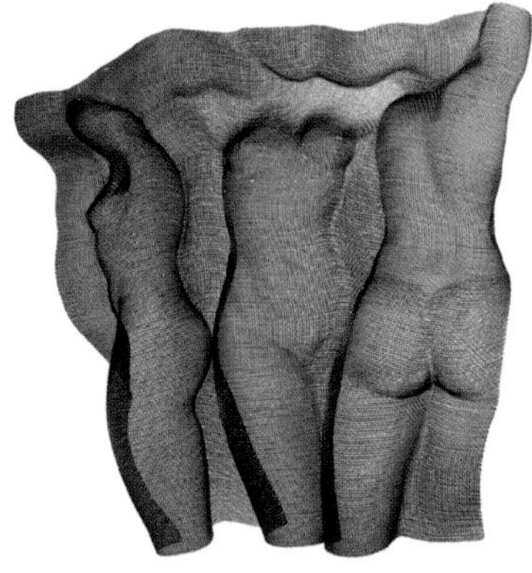

The Three Graces IX
Eric Boyer
Hand-formed steel wire
mesh sculpture
Courtesy the artist

Béla Julesz (1971). These are images of fields of random dots that have minimal form information when viewed in isolation, but are computationally manipulated to provide the percept of pure 3D surface image structure when a pair of the images are viewed appropriately with the two eyes. However, a key aspect of the resulting structures is that they are perceived as pure surfaces with no volumetric information. Thus, they do not have the sense of solidity normally associated with objects, but they seem more like traces of objects than the objects themselves, the transparent veils of the object surface structure that delineate the object without being the object [see figures left].

In this sense the interfaces represent a form of null object that is reminiscent of the early theory of how we see objects from Democritus and Lucretius, the early Greek and later Roman philosophers, respectively. They viewed the vision of objects in terms of thin skins, or "species", being continually emitted from every object and entering the eye to convey the information of the object's shape to the viewer. This remarkable viewpoint seems at first sight highly implausible and indeed was met with substantial philosophical scepticism. How could objects continually emit part of their surface? How could the emissions travel in all directions? How could they become small enough to enter the eye? And yet, Democritus' view almost exactly mirrors the current concept of light propagation as characterised by the representation of its optical wavefront emanating from the source object. The wavefront from reflected light contains the shape of the object and is, indeed, continually emitted in all directions. The wavefront is, moreover, the core of the wave equation that forms the basis of modern Quantum Theory. Within the eye, the optical wavefront is readily captured by the pupil at any point in space and retains its shape through the inverse optical transformation at the location of the nodal point (although this 3D form is not directly accessible within the eye). Cast in terms of wavefronts, the 'skins' proposed by Democritus and Lucretius seems to offer an accurate metaphor of the multifarious nature of reflected light as a carrier of information about objects. They are, nevertheless, an evanescent representation lacking solidity or manipulability, and therefore qualify as one form of null object that is specific to the object yet is not the object as such.

A further form of nullity with respect to stereoscopic object representation is the concept of negative depth or a 'hole' in the defined stereoscopic space. Such holes can be defined in two

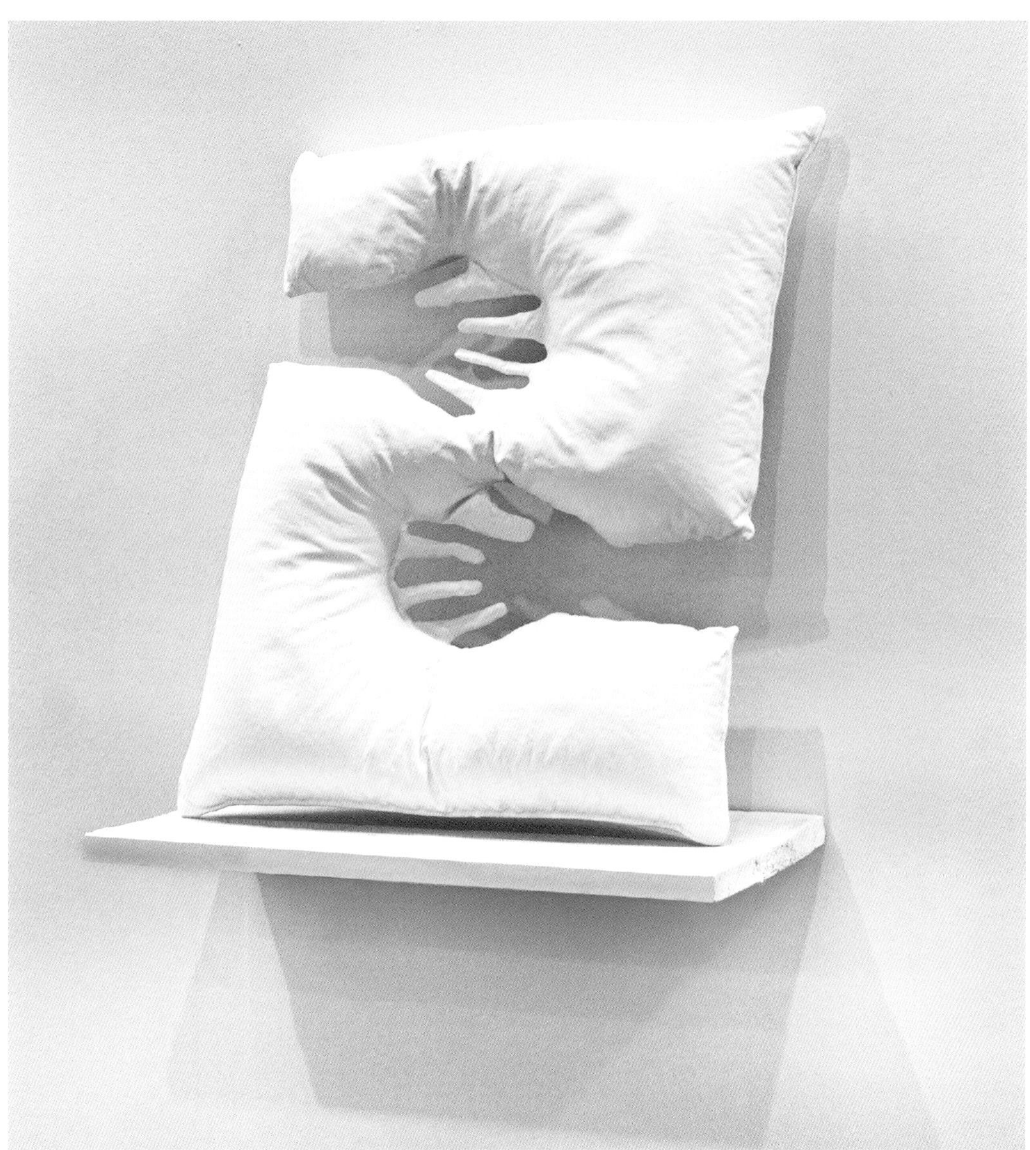

empreinte #1
Mathilde Roussel, 2011
Ceramic
24 x 18 x 4 in
Photography Mathilde Roussel
Courtesy the artist

ways, either as a negative form in the stereoscopic surface defined by the random dots, or as a negative space in a 3D array or volume specified by the stereoscopic disparity assigned to each dot. The negative form in the stereoscopic surface is readily achieved simply by switching the eye assignments of a stereoscopic pair defining a positive surface shape. One compelling form of this is the inverse form of a human body, or null body space, exemplified by the negative space pillow of Mathilde Roussel. This is a form that we never see in everyday life, since bodies are solid material and are

TOP

Rubin's vase
Illustration by London Fieldworks
2012

BOTTOM

Inward outward arrows illusion
Illustration by London Fieldworks
2012

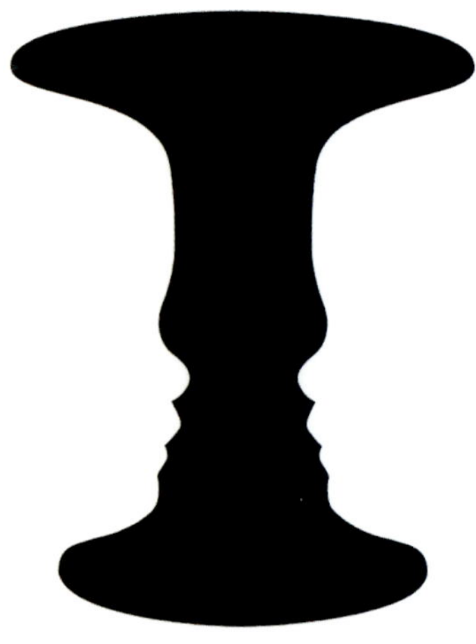

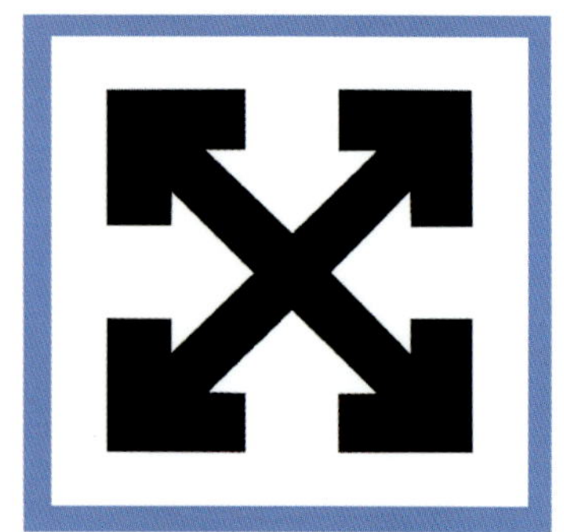

always viewed externally. They may be seen as physical art objects, such as the glass sculptures of NFN Kalyan or the wireframe sculptures of Eric Boyer, when viewed from behind. In both cases, these sculptures contain the stereoscopic cues of the binocular disparity defining the inverted depth image, just as in the case of the stereogram, but they also contain monocular depth cues that allow the form to be seen in the printed images, so they are not as pure in this sense as the random-dot stereogram, which contains no associated monocular depth cues.

In viewing such images, one can ask whether the viewers apprehend it as the negative space left by the absence of the object (i.e. an explicit form of null object), as the negative 3D form of the object (i.e. the inverse of the object, but not null in the strict sense), as the positive form of the object negatively viewed (i.e. from the inside, as though the material were transparent), or even with a perceptual flip as the positive form of the object, in the way that Gregory's 'hollow face' illusion inverts the negative 3D mask of the face to see it in its usual depth configuration (Króliczak, Heard, Goodale and Gregory, 2006). As the figures show, the last option is not one that is experienced, which is predictable from the fact that adding stereoscopic cues to the hollow face mask forces it to be seen in its true hollow form, and no longer allows the false inverted percept of a normal face.

The inverse representation of negative space is a widespread visualisation technique used in painting and sculpture, extensively used to represent the absence of specific objects as well as the structure of the spatial complement of positive objects. Here they are speaking not of negative depth in the sense of hole, but of the space and the shapes formed by the space around the objects that are the focus of attention. This idea is closely related to the Figure/ Ground Effect, in that the negative space is the unattended ground, which becomes a sort of inverse figure when attention is drawn to it. The illustrated examples are chosen over typical figure/ground examples because they are difficult to invert, with the initial impression maintaining its dominant status and emphasising the negative, subordinate role of the negative space. The term is apt for two further reasons also. One is that it is 'space' rather than 'object' in relation to the border ownership. Visual objects 'own' their borders, in a technical sense that has become widely accepted in perceptual studies (Koffka 1935; Nakayama and Shimojo, 1990; Zhou, Friedman and von der Heydt, 2000; Bertamini, 2006). Being a figure means that the contours that surround the figure

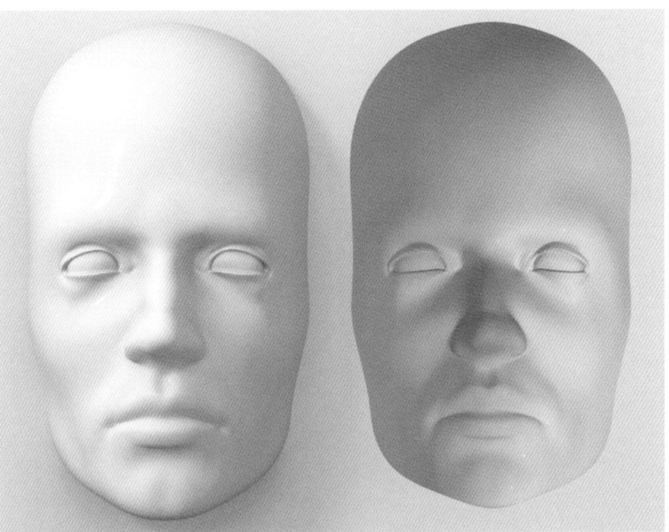

Mask and hollow mask
Photography David Mack
Science Photo Library

are not shared but are owned by the figure alone. This is not just
an abstract formalism but implies that the region that does not
own the border, the ground, has no border, and hence is seen as
a continuous field that runs behind the figure. Calling it 'negative
space' implies that, when the ground is attended, it acquires figural
properties to some degree, including the border ownership, and
that what was previously figure loses the borders to become the
continuous ground behind the negative space. These concepts are
illustrated by perusal of the three examples here.

One study of figure ground effects has focused on the brain
signals corresponding to the ground experience. In a functional
imaging investigation the cortical network for figure/ground
organisation, Likova and Tyler (2007) found that the only change
in the visual representation areas was a reduction in signal
strength in the regions representing the ground in the early visual
areas (V1 and V2), although there was no measurable change in
the regions representing the figure. This study was performed
by an instantaneous texture replacement paradigm designed to
be sensitive only to the level of perceptual organisation of the
stimulus information. Two other regions of cortex seemed to
be incorporated in the figure/ground network controlling the
perceptual organisation, but they exhibited no suppression. This
study therefore indicates that the 'ground' status of non-figural
regions is mediated by specific suppression of the neural activity
at the earliest levels of the cortical hierarchy for visual processing,
accounting for its nebulous perceptual experience.

In a sense, what has been discussed in the foregoing has focused on all the interpretations that are not truly null objects. They are negative to the concept of definite objects, but (contrary to Aristotle!) every concept does not have just one negative. That would be true in a one-dimensional world; going down is the opposite of going up. But the domain of conceptualisation is hyperdimensional, so not going in one direction leaves the opportunity of going in many other directions. Similarly, as we have seen, there are many ways to not perceive an object. However, the inchoate representation of the null object has in principle no form, 3D structure or content. It is null space in the core sense that collaborator Gustav Metzger is projecting, which seems to be the true absence of form that evokes the concept of empty formless void that is inherently contradictory to the concept of its realisation.

Another form of object nullity is the well-known difficulty of not thinking about a specified object: "Don't think about pink elephants." The first response to such a suggestion is, paradoxically, immediately to think about them. A more practical approach to achieve the desired result is to think about some other objects, such as green bicycle, but this does not seem to be a true polar opposite to thinking about pink elephants. In neural terms, we can think of the brain as containing a vast array of object representations, of which the pink elephant representation would be one local circuit. The true null case of thinking about pink elephants would be the local inhibition of this local circuit, such that even if one tried to think about them, nothing would be activated. This is more like the 'tip of the tongue' phenomenon in which, when you are trying to come up with a word or a concept, a mental blank is experienced. Here there is a meaningful null experience, a sense of hollow emptiness when attempting to access the concept, that can be more adequately described as the null case of this particular object. In an attempt to elicit this experience, try thinking of the word 'tope'. You may have once encountered it, perhaps in the context of the term 'isotope', but (unless you are a radiochemist) do you now recall the meaning of the term? If not, you probably have a sense that it points to a concept that has a meaning, but are faced with the empty mental space of what the specific meaning actually may be.

A strong version of this kind of blank-out is often experienced by marijuana smokers. The normal experience is that we have a clear idea of what we are going to say throughout a conversation, with the occasional loss of a word here or there to accurately describe the concept to be voiced at any point in time, but clear knowledge of the intended meaning of the word being sought and the following thought. In other words, we generally have a sense of a filled mental space. The experience of the marijuana smoker, on the other hand, can be of the complete loss of the contents of mental space partway through a sentence. It is like looking into a mental opacity that is emitting no information, a null space, or as though the horse one was riding suddenly disappears from under one's seat. In terms of current neuroscience understanding, the neural substrate of this form of blank-out has not been resolved, but the leading candidate would be the hippocampus, which plays a key role in coordinating the operation of working memory, or consciousness of items retrieved from long term memory for use in a conversation. If this is the site of working memory awareness, it would have to be presumed that the activity in the hippocampus would be suddenly suspended at the moment of the loss of the train of thought. This is not a loss of consciousness but the vivid consciousness of the momentary loss of mental contents.

The illustration shows the tubular form of the hippocampus (one on each side of the brain), connecting up to the looping structures of the limbic system that encodes our emotional responses. These neuroanatomical connections explain why emotions play such a large part in memory retention. Not so easily seen is the fact that the hippocampus is directly connected along its length to the nearby cortex, which is the long-term memory storehouse on which the hippocampus is operating.

Both the mystery of the null image and the emotions it can evoke are expressed by Leonardo da Vinci in a description of a boyhood experience in the mountains in *Codex Arundel*, c. 1480:

> I wandered some way amongst the shadowy rocks and eventually reached the mouth of a huge cavern before which I stood a while, disoriented by the place I had never encountered before. I stooped over with arched back, my left hand on one knee, my right hand shading my furrowed brow; and twisting this way and that I peered in to try to make out if anything was inside, but the deep darkness prevented me from doing so. After being there for some time, two things suddenly arose in me—fear and desire; fear of that dark, threatening cave, and desire to see if there was some marvelous thing inside.

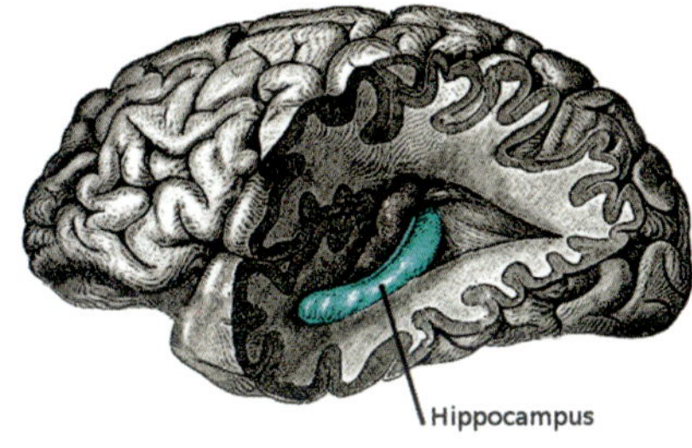

Hippocampus
Modified from a scan of a plate of "Posterior and inferior cornua of left lateral ventricle exposed from the side" in *Gray's Anatomy* by Looie496, 2008

The same spirit of the exploration of the darkness is captured by Nietzsche's quote: "When you stare into the abyss the abyss stares back at you." This brings to mind the image of the eye itself as a dark chamber, or camera obscura, developed to capture images of the world outside with maximum efficiency. When we stare in to that dark chamber we have a sense of probing the secrets of the consciousness behind it, as illustrated in the close up of the pupil of an eye at the beginning of this essay. As we stare into that blackness, it seems to come alive with a shimmering, scintillating energy. Indeed this perceptual energy has been used for centuries to open up the channels of mental (they would say 'psychic') creativity in the form of the ultrablack 'scrying mirror' into which adepts would stare to derive images that might be interpreted as predicting the future. In physics terms, we are seeing the actual quantum energy of the light reception in our own eyes, which are sensitive enough to detect single quantum detection events in the millions of photopigment molecules in each photoreceptor at the back of our eyes. We are, in fact, directly in touch with the nanoscale quantum events making up the energetics of universal activity, revealed in our perceptual experiences.

Perhaps this same dark energy is what Kazimir Malevich was thinking of when he painted his *Black Square*, 1915. Did he intend

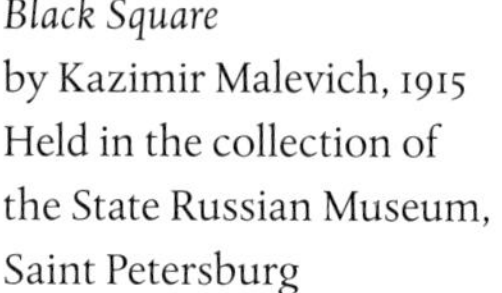

Black Square
by Kazimir Malevich, 1915
Held in the collection of
the State Russian Museum,
Saint Petersburg

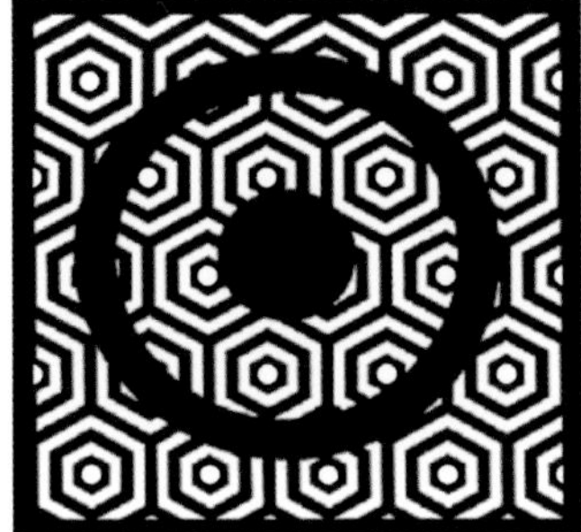 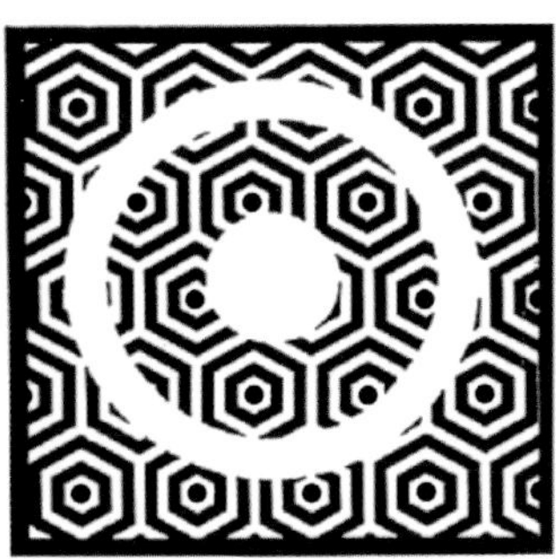

Contrast illusion
llustration by
London Fieldworks, 2012
after Jacques Ninio, 2004

a reference to the dynamic energy of the visual experience of blackness, the null point energy of quantum physics or the returning stare of Nietzsche (all of which were in the air by the time he was painting). The effect may be easier to see under the original studio lighting than in a brightly lit gallery. Ironically, as time has passed the work has acquired a form of painterly energy in the form of the 'craquelure' of the surface into a cascade of dynamic structure that we can make out if we peer into its blackness. As its shape coalesces, could it be channelling Schrödinger's Cat?

This blackness scintillation effect is captured to some extent by the figure from Ninio, 2004, emphasising the 'holeness' of black as opposed to the 'surfaceness' of white. Especially if viewed on a high-blackness computer screen, staring at the center of the black region will reveal not only that it appears to be a hole rather than a surface, but a tendency to experience scintillations of some kind, as though the blackness has a life of its own.

My own research has investigated the neural basis of these induced scintillations in a functional imaging study. The stimulus was a flickering windmill checkerboard on a grey background, following by a null period with no stimulation. Just as in the dark pupil, Malevich and Ninio images, the high contrast regions tend to induce scintillations in the intervening grey regions, which are most clearly seen when viewing a uniform grey screen during the

Stimulation conditions
for the induced twinkle effect
by Chen, Tyler, Liu and Wang,
2005
Image courtesy
Christopher Tyler

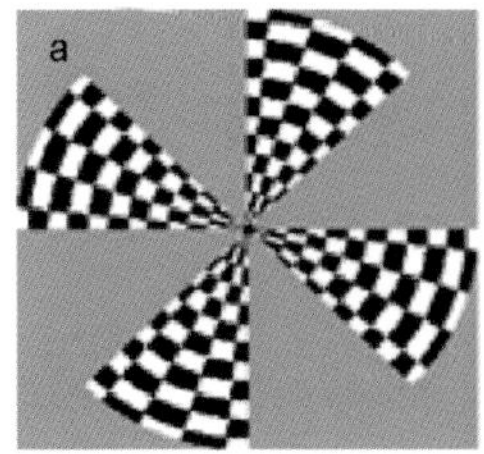 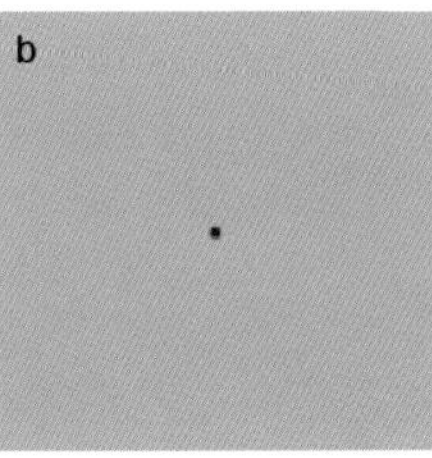 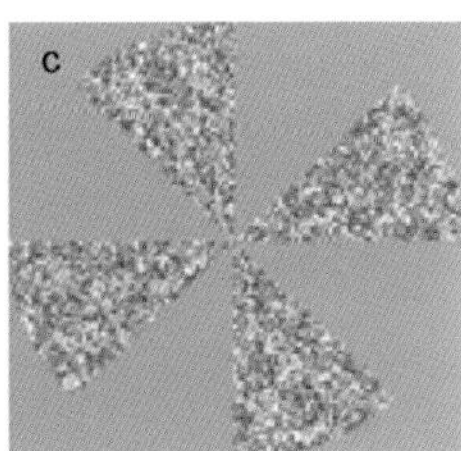

null period (Tyler and Hardage, 1994). Thus, the induced twinkle, which persists for many seconds after removal of the flickering regions, is a null effect being perceived in regions of the eye that have never received any stimulation. Are these just derivative of some kind of cognitive construct or do they have a specific basis in the visual processing stream in the cortex?

By imaging the brain while undergoing the induced twinkle aftereffect, Chen, Tyler, Liu and Wang (2005) were able to establish that the induction occurred in the primary visual cortex, presumably by long range connections released from inhibition as the stimulus was removed. The remarkable aspect, however, is that the resultant release is able to generate the dynamic, twinkle effect in the unstimulated cortical regions which, by virtue of its chaotic noise spectrum, contains the possibility of endless images that may stimulate the recognition of objects from the null field. In this way, the blankness of the absence of objects is turned into genuine activation that, in turn, stimulates the awareness of objects from the hyperdimensional space of the memory/imagination that derives from the accumulation of object representations from our lifetime of experience.

Reprise: close-up of the blackness of the human stare
Modified from Look into
My Eyes, 2012

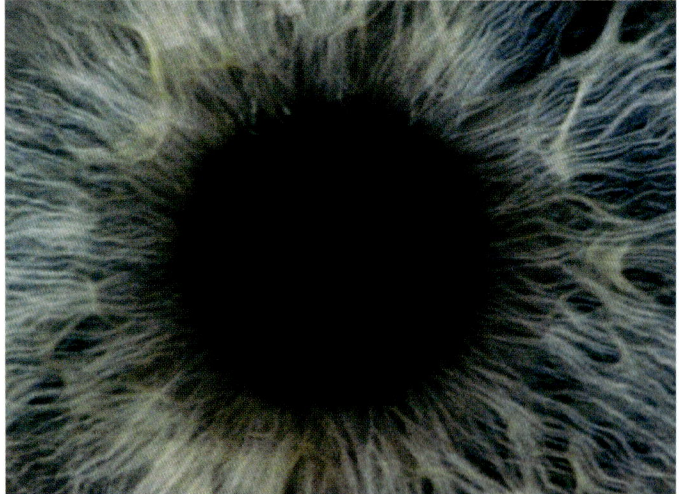

REFERENCES

Adelson, Edward H, "On seeing stuff: The perception of materials by humans and machines", *Proceedings of the SPIE Vol. 4299, Human Vision and Electronic Imaging VI*, 2001, pp. 1–12.

Bertamini, Marco, "Who owns the contour of a visual hole?", *Perception*, vol. 35, 2006, pp. 883–894.

Chen, CC, with CW Tyler, CL Liu, YH Wang, "Lateral modulation of BOLD activation in unstimulated regions of the human visual cortex", *NeuroImage 24*, 2005, pp. 802–809.

Koffka, K, *Principles of Gestalt Psychology*, London : Lund Humphries, 1935.

Króliczak, G, with MA Goodale, RL Gregory, P Heard, "Dissociation of perception and action unmasked by the hollow-face illusion", *Brain Res. 1080*, 2006, pp. 9–16.

Likova, LT, CW Tyler, "Occipital network for figure/ground organization", *Experimental Brain Research 189*, pp. 257–267.

Nakayama, K, S Shimojo, "Toward a neural understanding of visual surface representation", *Cold Spring Harb Symp Quant Biol.* 55, 1990, pp. 911–924.

Ninio, J, "Scintillations, extinctions, and other new visual effects", in A Carsetti ed., *Seeing, Thinking and Knowing*, New York : Springer-Kluwer, 2004.

Tyler CW, with L Hardage, "Induced twinkle aftereffect as a probe of visual processing mechanisms", *Vision Res.* 35, 1994, pp. 757–776.

Zhou H, with HS Friedman, R von der Heydt, "Coding of border ownership in monkey visual cortex", *J Neurosci.* 20, 2000, pp. 6594–6611.

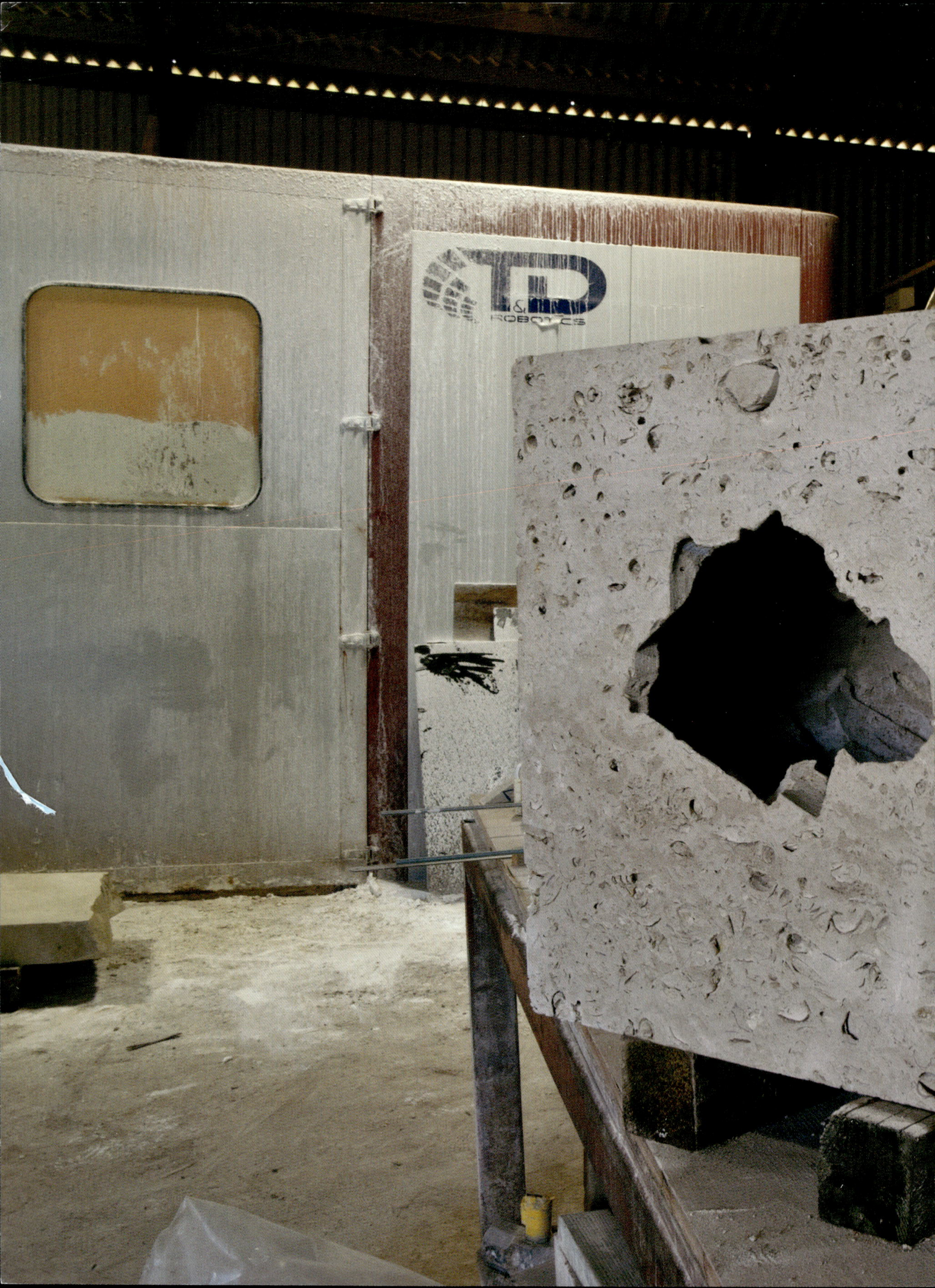

T&D
ROBOTICS

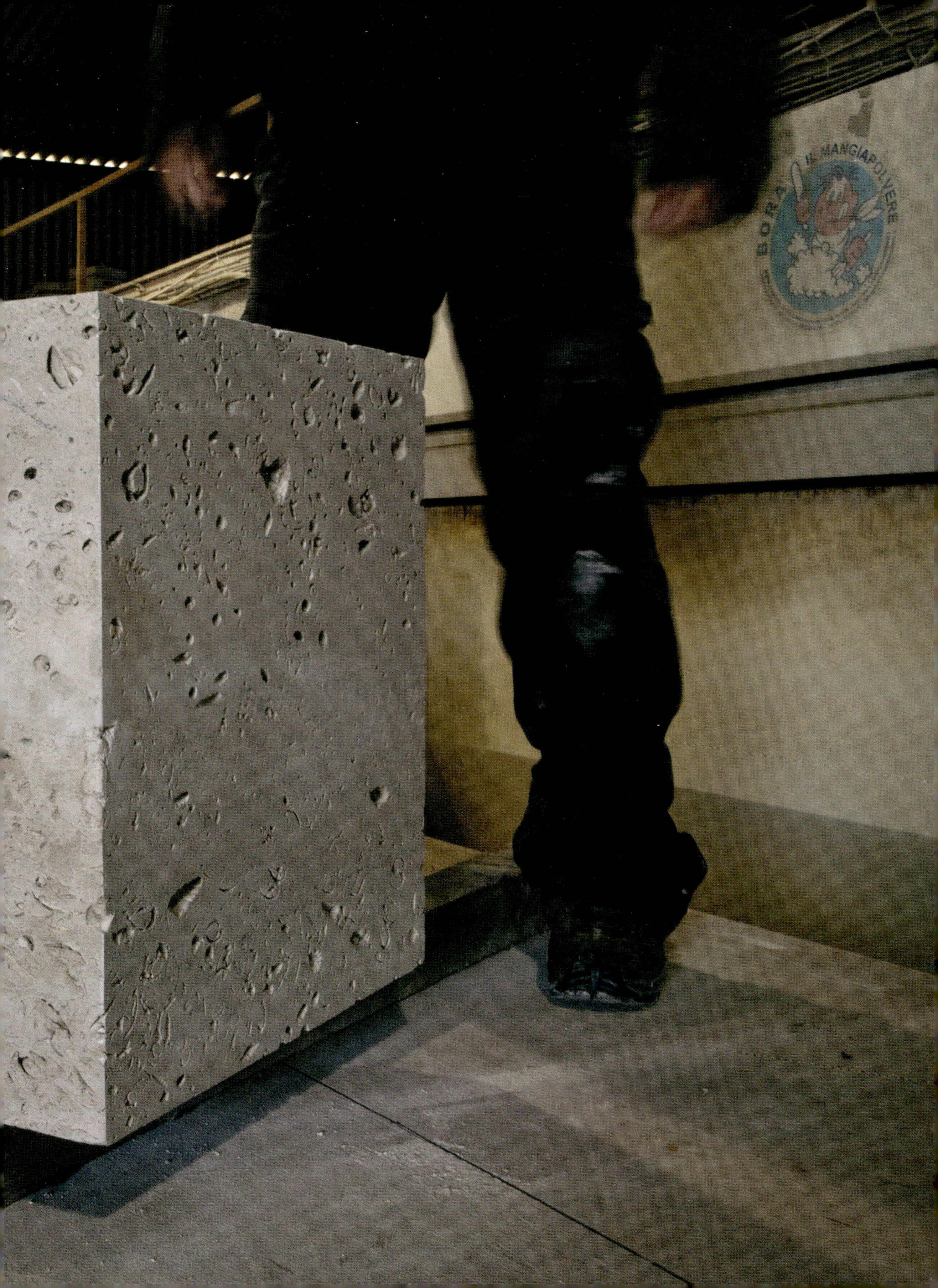
BORA IL MANGIAPOLVERE

EMPTY AND EMPTINESS, AND EMPTYING

Gustav Metzger

The Sg[1] is particularly well suited: it consists of a series of interlocking rooms, surrounded by a large park with trees.

Treading lightly—*Man ist verunsichert.* Difference between empty and emptiness, and emptying? Vide: we remove the rug from our feet. Expectations not fulfilled. In this emptied space, we place matter that on the one hand gels and on the other frictionalises. We play with space. Here we can recall Martin Creed's balloon installation. Anthony Gormley's Hayward Gallery installation is a further reference point.

The emptied space is squeezed by the inserted work. The work becomes an insertion from the stand point—from the point of view of emptied space—alien even here we can clearly see why any texts or pieces of paper introduce disharmony. Turbine Hall. A new methodology.

... tiptoeinthewillow... com

Verunsicherung has to be choreographed—We submit an invisible choreography. An entirely new way of configurating an exhibition.

The increasing complexification of life, coupled with increasing study and analysis of reality, obliges us in the visual art field to reassess procedures.

What we are after is a more synthesised interaction between audience and the art object making up an exhibition—what can we offer? Essentially this consists of the placing of objects within the gallery space.

This is done as if the space was empty.

But what if the space is emptied?

An alien insertion—the holy. Eastern. Several layers of *verunsicherung.*

Normal procedures are how to fill it. We now have to give more attention to the surrounding space, the emptied gallery, the gallery that is fulfilled by its emptiness, demands exceptional care from curators. First we visualise an emptied space. We now see that space entered into by specific configurations. Does it matter what these figurations consist of—what they mean?

From the point of view of the emptied space, the configurations are no more than a nuisance—to be got rid of at the earliest opportunity: it is alien, serves no purpose.

How should we now react? We don't want to force the issue. How can we accommodate the empty space? Make it feel content? The first and most obvious answer is to bring in as little as possible. Squeeze the matter and so gain "space" for emptiness. And why should this not be done?

Once we concede that emptiness is a plus we are obliged to draw a range of conclusions whose range could threaten our entire art-world. If a feasible scenario proposed by a series of thinkers that the best way out for nature as a whole is the self abolition of humanity, to allow the continuation of the nature, the radical consequences of emptiness becomes less threatening and less alien.

And so, in curating, and indeed in making art, another sensibility is called for, one that accepts the freedom of the other, that is the range of demands that emptiness may make be considered at every stage of exhibition making.

In the process, we can find that new planes of understanding open up.

In the case of Creed, space becomes a trap, a surge of danger. But transfer this uninhibited stampede to the calm atmosphere of the empty gallery?

This is where we must again question a range of habitual activities.

[1] Sg = Serpentine gallery

CONTRIBUTORS

Bronac Ferran is a London-based, Irish-born curator, researcher and writer. She specialises in works hovering at the intersections between disciplines as is outlined at her website www.boundaryobject.org. Her most recent exhibition was Poetry, Language, Code at the Ruskin Gallery in Cambridge in June 2012, where she was guest curator. She is a reviewer for *Neural* magazine and has written for recent exhibition catalogues on the work of Gina Czarnecki and Fernando Velázquez as well as *Leonardo Journal of Art and Science*. She was on the jury for the Ars Electronica Hybrid Arts award in 2010 and 2011 and for the transmediale Award in Berlin in 2009. She is a former Director of Interdisciplinary Arts at Arts Council England where she was Executive Commissioner with Tony White of the *Pioneers in Art and Science* DVD, made by Ken McMullen with Gustav Metzger, and set up numerous initiatives in art and ecology, law and science. She now works part time at the Royal College of Art.

Hari Kunzru is the author of the novels *The Impressionist*, 2002, *Transmission*, 2004, *My Revolutions*, 2007, and *Gods Without Men*, 2011, as well as a short story collection, *Noise*, 2006. His work has been translated into twenty-one languages and won prizes including the Somerset Maugham award, the Betty Trask prize of the Society of Authors, a Pushcart prize and a British Book Award. In 2003 Granta named him one of its 20 best young British novelists. *Lire* magazine named him one of its 50 "*écrivains pour demain*". He is Deputy President of English PEN, a patron of the Refugee Council and a member of the editorial board of *Mute* magazine. His short stories and journalism have appeared in diverse publications including *The New York Times, Guardian, New Yorker, Financial Times, Times of India, Wired* and *New Statesman*. He lives in New York City.

Dr Nick Lambert is Lecturer in Digital Art and Culture at Birkbeck, University of London, in the Department of History of Art and Screen Media. He teaches in the area of computer-based art, the history of technological art, digital curation and media conservation. Lambert is also active in other institutions, including Ravensbourne College of Art, where he teaches on the BA and MA contextual studies courses. He obtained his DPhil at Oxford University in 2003, where his supervisor was Professor Martin Kemp. Lambert was formerly the Research Fellow for the AHRC Computer Arts, Contexts, Histories, etc (CACHe) Project, and Principal Investigator on the Computer Art and Technocultures (CAT) Project. He maintains an active role as Chair of the Computer Arts Society and Exhibition Chair of the EVA Conference. He also develops digital art projects, including Fulldome projections and installations, and his works Oculus, *Lux Nova* and *Music of the Spheres* have been exhibited in the UK, USA, and elsewhere.

London Fieldworks was formed in 2000 by artists Bruce Gilchrist and Jo Joelson as a cross-disciplinary, collaborative practice working across social engagement, architecture, sculpture, installation and video, situating works in the gallery and the landscape. Early works interrogated ideas around authenticity of mediated experience, especially experience of place, while exploring the site-specific nature of cognition. These projects were seminal to the artists' notion of ecology as a complex inter-working of social, natural, and technological worlds. Recent projects have explored concepts of 'performative architecture' in rural and urban green spaces and include *Super Kingdom*, 2008, *Outlandia*, 2010, and *Spontaneous City*, 2010–2012. In 2009, as recipients of a British Council residency, they worked in the Atlantic Rainforest in Brazil. Their work has featured in publications including *Far Field*, 2011; *Searching for Art's New Publics*, 2010; *ART+SCIENCE NOW*, a visual survey of artists working at the frontiers of science and technology, 2010; *Beyond Architecture: Imaginative Buildings and Fictional Cities*, 2009. Their site-specific installation, *Super Kingdom* and video-animation, *Monarchy* was awarded an Honorary Mention in the Hybrid Art category of Ars Electronica in 2010.

Gustav Metzger is a London-based artist, born in Nuremberg, Germany in 1926 to Polish-Jewish parents. He was evacuated to England with his brother, Max/Mendel as part of the Kindertransport in 1939. From 1945 to 1953, Metzger studied at various art schools in Cambridge, London, including David Bomberg's class at the Borough Polytechnic, Antwerp and Oxford. In 1959, he developed the concept of 'Auto-destructive art', proposing works that could self-destruct, to reflect the similarly destructive nature of political and social systems. At the heart of his practice, which has spanned over 60 years, are a series of constantly opposing yet interdependent forces such as destruction and creation.

Dr Christopher W Tyler's research career is in visual perception and visual neuroscience, focused on theoretical, psychophysical, oculomotor and functional MRI studies of the neural substrates of line, pattern, symmetry and stereoscopic depth perception. He has a longstanding interest in the interface between art and the scientific study of visual perception. His art interests include portraiture and general principles of composition, the history of perspective techniques. In the mid-1970s, his work was included in one of the early compendia of computer art. His subsequent invention of random-dot single-image ('Magic Eye') autostereograms for viewing printed 3D forms made specific reference to concepts from Leonardo, Escher, Lichtenstein, Vasarely and others. More recently, Tyler discovered a general principle of portrait composition that one eye is placed close to the centre line of the canvas in portraits throughout history. He was one of the few scientists invited to debunk David Hockney's book *Secret Knowledge*, proposing the early use of optics by artists. He gives numerous lectures at art institutes and universities around the world.

ACKNOWLEDGEMENTS

Thanks to Bronac Ferran and Nick Lambert for connecting us to the Computer Arts Society and supporting the project from the outset.

Our profound thanks to Gustav Metzger for donating his thoughts; also for permissions to publish photographs of his artworks and the text "Empty and Emptiness, and Emptying".

Thanks are due to those who have made the crucial connections between people, material and machines. These include programmer Jonny Bradley who wrote the source code to connect Gustav with the database to create the primitive shape information. David Noonan for introducing us to the robotic manufacturing sector, Yaroslav Tenzer who facilitated the CAD-CAM process and Steven Newbury who carried out the testing and the machining process with the KUKA robot at Stoneworld.

For the insightful and unique perspectives they have provided in writing texts for this book, we are grateful to Bronac Ferran, Hari Kunzru, Nick Lambert and Christopher Tyler.

We would like to thank Irini Papadamitriou and the Victoria and Albert Museum for the opportunity to present *Looking at Primitives* as part of the Digital Weekend 2012.

We are grateful to Ula Dajerling and Leanne Dmyterko (assistants to Gustav) for helping to arrange access to Gustav's paintings and for providing Gustav's biog information for the contributor's section.

We would like to thank Mathieu Copeland for forwarding the Voids: A Retrospective catalogue and Gustav's text "Empty and Emptiness, and Emptying".

Our thanks are due also to Adrian Glew, Archivist at Tate Britain who kindly located the early stone carving work in Gustav Metzger's personal archive held there.

Special thanks are due to Black Dog and WORK gallery for supporting this project and publication.

We gratefully acknowledge the financial assistance of the Arts Council of England and the Computer Arts Society.

CREDITS

Meyrink, Gustav, *The Golem*, 1915, translation by Mike Mitchell, published by Dedalus, 1995, p. 87.
All images courtesy of London Fieldworks except where stated otherwise.

COLOPHON

Copyright 2012 Black Dog Publishing Limited,
London, United Kingdom and the authors.
All rights reserved.

Edited by Bruce Gilchrist & Jo Joelson.
Designed by Laura Varžalgytė at
Black Dog Publishing Limited.

Black Dog Publishing Limited
10a Acton Street, London WC1X 9NG
United Kingdom

Tel: +44 (0)20 7713 5097
Fax: +44 (0)20 7713 8682
info@blackdogonline.com
www.blackdogonline.com

British Library Cataloguing-in-Publication Data.
A CIP record for this book is available from the
British Library.

ISBN 978 1 908966 12 4

Black Dog Publishing Limited, London, UK,
is an environmentally responsible company.

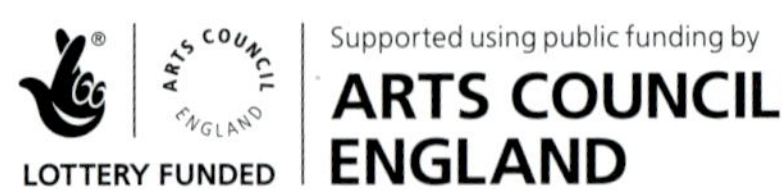

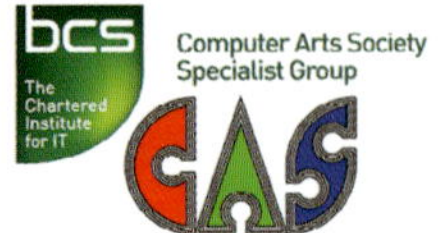

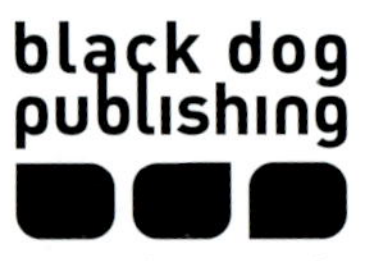

art design fashion
history photography
theory and things

www.blackdogonline.com

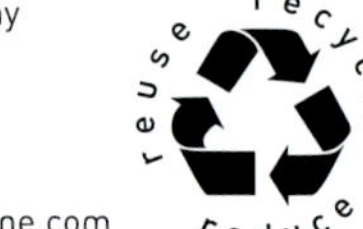

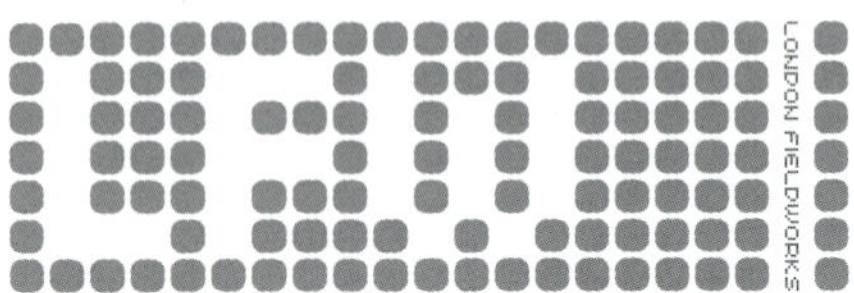

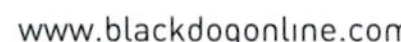